Chip Multiprocessor Architecture: Techniques to Improve Throughput and Latency

Chip Multiprocessor Architecture: Techniques to Improve Throughput and Latency

Kunle Olukotun, Lance Hammond, and James Laudon

www.morganclaypool.com

ISBN: 159829122X   paperback
ISBN: 9781598291223   paperback

ISBN: 1598291238   ebook
ISBN: 9781598291230   ebook

DOI: 10.2200/S00093ED1V01Y200707CAC003

A Publication in the Morgan & Claypool Publishers series
SYNTHESIS LECTURES ON COMPUTER ARCHITECTURE #3

Lecture #3
Series Editor: Mark D. Hill, University of Wisconsin

Library of Congress Cataloging-in-Publication Data

Series ISSN: 1935-3235   print
Series ISSN: 1935-3243   electronic

First Edition
10 9 8 7 6 5 4 3 2 1

Synthesis Lectures on Computer Architecture

Editor
Mark D. Hill, *University of Wisconsin, Madison*

Synthesis Lectures on Computer Architecture publishes 50- to 150 page publications on topics pertaining to the science and art of designing, analyzing, selecting and interconnecting hardware components to create computers that meet functional, performance and cost goals.

Chip Mutiprocessor Architecture: Techniques to Improve Throughput and Latency
Kunle Olukotun, Lance Hammond, James Laudon
2007

Transactional Memory
James R. Larus, Ravi Rajwar
2007

Quantum Computing for Computer Architects
Tzvetan S. Metodi, Frederic T. Chong
2006

Chip Multiprocessor Architecture: Techniques to Improve Throughput and Latency

Kunle Olukotun
Stanford University

Lance Hammond
Stanford University

James Laudon
Sun Microsystems

SYNTHESIS LECTURES ON COMPUTER ARCHITECTURE #3

MORGAN & CLAYPOOL PUBLISHERS

ABSTRACT

Chip multiprocessors — also called multi-core microprocessors or CMPs for short — are now the only way to build high-performance microprocessors, for a variety of reasons. Large uniprocessors are no longer scaling in performance, because it is only possible to extract a limited amount of parallelism from a typical instruction stream using conventional superscalar instruction issue techniques. In addition, one cannot simply ratchet up the clock speed on today's processors, or the power dissipation will become prohibitive in all but water-cooled systems. Compounding these problems is the simple fact that with the immense numbers of transistors available on today's microprocessor chips, it is too costly to design and debug ever-larger processors every year or two.

CMPs avoid these problems by filling up a processor die with multiple, relatively simpler processor cores instead of just one huge core. The exact size of a CMP's cores can vary from very simple pipelines to moderately complex superscalar processors, but once a core has been selected the CMP's performance can easily scale across silicon process generations simply by stamping down more copies of the hard-to-design, high-speed processor core in each successive chip generation. In addition, parallel code execution, obtained by spreading multiple threads of execution across the various cores, can achieve significantly higher performance than would be possible using only a single core. While parallel threads are already common in many useful workloads, there are still important workloads that are hard to divide into parallel threads. The low inter-processor communication latency between the cores in a CMP helps make a much wider range of applications viable candidates for parallel execution than was possible with conventional, multi-chip multiprocessors; nevertheless, limited parallelism in key applications is the main factor limiting acceptance of CMPs in some types of systems.

After a discussion of the basic pros and cons of CMPs when they are compared with conventional uniprocessors, this book examines how CMPs can best be designed to handle two radically different kinds of workloads that are likely to be used with a CMP: highly parallel, *throughput-sensitive* applications at one end of the spectrum, and less parallel, *latency-sensitive* applications at the other. Throughput-sensitive applications, such as server workloads that handle many independent transactions at once, require careful balancing of all parts of a CMP that can limit throughput, such as the individual cores, on-chip cache memory, and off-chip memory interfaces. Several studies and example systems, such as the Sun Niagara, that examine the necessary tradeoffs are presented here. In contrast, latency-sensitive applications — many desktop applications fall into this category — require a focus on reducing inter-core communication latency and applying techniques to help programmers divide their programs into multiple threads as easily as possible. This book discusses many techniques that can be used in CMPs to simplify parallel programming, with an emphasis on research directions proposed at Stanford University. To illustrate the advantages possible with a CMP using a couple of solid examples, extra focus is given to *thread-level speculation* (TLS), a way to automatically break up nominally sequential applications into parallel threads on a CMP, and *transactional memory*. This model can greatly simplify manual parallel programming by using hardware — instead of conventional software locks — to enforce atomic code execution of blocks of instructions, a technique that makes parallel coding much less error-prone.

KEYWORDS

Basic Terms: chip multiprocessors (CMPs), multi-core microprocessors, microprocessor power, parallel processing, threaded execution

Application Classes: throughput-sensitive applications, server applications, latency-sensitive applications, desktop applications, SPEC benchmarks, Java applications

Technologies: thread-level speculation (TLS), JRPM virtual machine, tracer for extracting speculative threads (TEST), transactional memory, transactional coherency and consistency (TCC), transactional lock removal (TLR)

System Names: DEC Piranha, Sun Niagara, Sun Niagara 2, Stanford Hydra

Contents

CHAPTER 1

The Case for CMPs

The performance of microprocessors that power modern computers has continued to increase exponentially over the years, as is depicted in Fig. 1.1 for Intel processors, for two main reasons. First, the transistors that are the heart of the circuits in all processors and memory chips have simply become faster over time on a course described by Moore's law [1], and this directly affects the performance of processors built with those transistors. Moreover, actual processor performance has increased faster than Moore's law would predict [2], because processor designers have been able to harness the increasing number of transistors available on modern chips to extract more parallelism from software.

An interesting aspect of this continual quest for more parallelism is that it has been pursued in a way that has been virtually invisible to software programmers. Since they were first

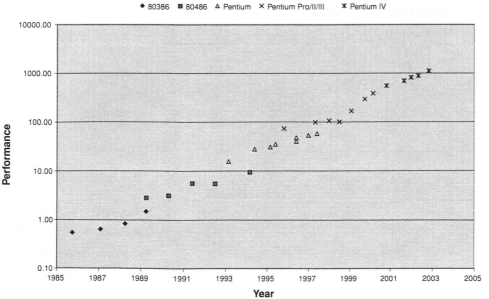

FIGURE 1.1: Intel processor performance over time, tracked by compiling published SPEC CPU figures from Intel and normalizing across varying suites (89, 92, 95, 2000).

invented in the 1970s, microprocessors have continued to implement the conventional Von Neumann computational model, with very few exceptions or modifications. To a programmer, each computer consists of a single processor executing a stream of sequential instructions and connected to a monolithic "memory" that holds all of the program's data. Because of the economic benefits of backward compatibility with earlier generations of processors are so strong, hardware designers have essentially been limited to enhancements that have maintained this abstraction for decades. On the memory side, this has resulted in processors with larger cache memories, to keep frequently accessed portions of the conceptual "memory" in small, fast memories that are physically closer to the processor, and large register files to hold more active data values in an extremely small, fast, and compiler-managed region of "memory." Within processors, this has resulted in a variety of modifications that are designed to perform one of two goals: increasing the number of instructions from the processor's instruction sequence that can be issued on every cycle, or increasing the clock frequency of the processor faster than Moore's law would normally allow. Pipelining of individual instruction execution into a sequence of stages has allowed designers to increase clock rates as instructions have been sliced into a larger number of increasingly small steps, which are designed to reduce the amount of logic that needs to switch during every clock cycle. Instructions that once took a few cycles to execute in the 1980s now often take 20 or more in today's leading-edge processors, allowing a nearly proportional increase in the possible clock rate. Meanwhile, superscalar processors were developed to execute multiple instructions from a single, conventional instruction stream on each cycle. These function by dynamically examining sets of instructions from the instruction stream to find ones capable of parallel execution on each cycle, and then executing them, often out-of-order with respect to the original sequence. This takes advantage of any parallelism that may exist among the numerous instructions that a processor executes, a concept known as *instruction-level parallelism* (ILP). Both pipelining and superscalar instruction issues have flourished because they allow instructions to execute more quickly while maintaining the key illusion for programmers that all instructions are actually being executed sequentially and in-order, instead of overlapped and out-of-order.

Unfortunately, it is becoming increasingly difficult for processor designers to continue using these techniques to enhance the speed of modern processors. Typical instruction streams have only a limited amount of usable parallelism among instructions [3], so superscalar processors that can issue more than about four instructions per cycle achieve very little additional benefit on most applications. Figure 1.2 shows how effective real Intel processors have been at extracting instruction parallelism over time. There is a flat region before instruction-level parallelism was pursued intensely, then a steep rise as parallelism was utilized usefully, and followed by a tapering off in recent years as the available parallelism has become fully exploited. Complicating matters further, building superscalar processor cores that can exploit more than a

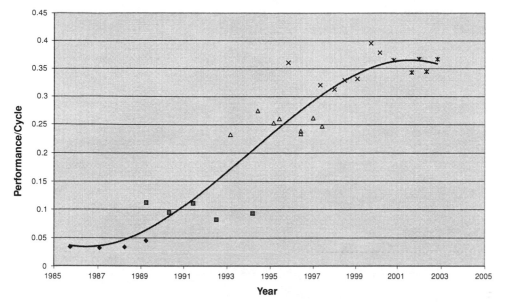

FIGURE 1.2: Intel processor normalized performance per cycle over time, calculated by combining intel's published (normalized) SPEC CPU figures and clock frequencies.

few instructions per cycle becomes very expensive, because the complexity of all the additional logic required to find parallel instructions dynamically is approximately proportional to the square of the number of instructions that can be issued simultaneously. Similarly, pipelining past about 10–20 stages is difficult because each pipeline stage becomes too short to perform even a basic logic operation, such as adding two integers together, and subdividing circuitry beyond this point is very complex. In addition, the circuitry overhead from adding additional pipeline registers and bypass path multiplexers to the existing logic combines with performance losses from events that cause pipeline state to be flushed, primarily branches, to overwhelm any potential performance gain from deeper pipelining after about 30 stages or so. Further advances in both superscalar issue and pipelining are also limited by the fact that they require ever-larger number of transistors to be integrated into the high-speed central logic within each processor core—so many, in fact, that few companies can afford to hire enough engineers to design and verify these processor cores in reasonable amounts of time. These trends slowed the advance in processor performance, but mostly forced smaller vendors to forsake the high-end processor business, as they could no longer afford to compete effectively.

Today, however, progress in processor core development has slowed dramatically because of a simple physical limit: *power*. As processors were pipelined and made increasingly superscalar over the course of the past two decades, typical high-end microprocessor power went from less than a watt to over 100 W. Even though each silicon process generation promised a reduction

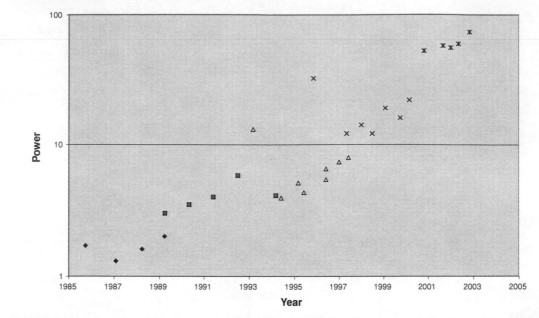

FIGURE 1.3: Intel processor power over time, calculated by combining published (normalized) SPEC and power numbers from Intel.

in power, as the ever-smaller transistors required less power to switch, this was only true in practice when existing designs were simply "shrunk" to use the new process technology. However, processor designers kept using more transistors in their cores to add pipelining and superscalar issue, and switching them at higher and higher frequencies, so the overall effect was that *exponentially more* power was required by each subsequent processor generation (as illustrated in Fig. 1.3). Unfortunately, cooling technology does not scale exponentially nearly as easily. As a result, processors went from needing no heat sinks in the 1980s, to moderate-sized heat sinks in the 1990s, to today's monstrous heat sinks, often with one or more dedicated fans to increase airflow over the processor. If these trends were to continue, the next generation of microprocessors would require very exotic cooling solutions, such as dedicated water cooling, that are economically impractical in all but the most expensive systems.

The combination of finite instruction parallelism suitable for superscalar issue, practical limits to pipelining, and a "power ceiling" set by practical cooling limitations limits future speed increases within conventional processor cores to the basic Moore's law improvement rate of the underlying transistors. While larger cache memories will continue to improve performance somewhat, by speeding access to the single "memory" in the conventional model, the simple fact is that without more radical changes in processor design, one can expect that microprocessor performance increases will slow dramatically in the future, unless processor designers find new ways to *effectively* utilize the increasing transistor budgets in high-end silicon chips to improve performance in ways that minimize both additional power usage and design complexity.

1.1   A NEW APPROACH: THE CHIP MULTIPROCESSOR (CMP)

These limits have combined to create a situation where ever-larger and faster uniprocessors are simply impossible to build. In response, processor manufacturers are now switching to a new microprocessor design paradigm: the chip multiprocessor, or CMP. While we generally use this term in this book, it is also synonymous with "multicore microprocessor," which is more commonly used by industry. (Some also use the more specific term "manycore microprocessor" to describe a CMP made up of a fairly large number of very simple cores, such as the CMPs that we discuss in more detail in Chapter 2, but this is less prevalent.) As the name implies, a chip multiprocessor is simply a group of uniprocessors integrated onto the same processor chip so that they may act as a team, filling the area that would have originally been filled with a single, large processor with several smaller "cores," instead.

CMPs require a more modest engineering effort for each generation of processors, since each member of a family of processors just requires stamping down a number of copies of the core processor and then making some modifications to relatively slow logic connecting the processors to tailor the bandwidth and latency of the interconnect with the demands of the processors—but does not necessarily require a complete redesign of the high-speed processor pipeline logic. Moreover, unlike with conventional multiprocessors with one processor core per chip package, the system board design typically only needs minor tweaks from generation to generation, since externally a CMP looks essentially the same from generation to generation, even as the number of cores within it increases. The only real difference is that the board will need to deal with higher memory and I/O bandwidth requirements as the CMPs scale, and slowly change to accommodate new I/O standards as they appear. Over several silicon process generations, the savings in engineering costs can be very significant, because it is relatively easy to simply stamp down a few more cores each time. Also, the same engineering effort can be amortized across a large family of related processors. Simply varying the numbers and clock frequencies of processors can allow essentially the same hardware to function at many different price and performance points.

Of course, since the separate processors on a CMP are visible to programmers as separate entities, we have replaced the conventional Von Neumann computational model for programmers with a new *parallel programming model*. With this kind of model, programmers must divide up their applications into semi-independent parts, or *threads*, that can operate simultaneously across the processors within a system, or their programs will not be able to take advantage of the processing power inherent in the CMP's design. Once threading has been performed, programs can take advantage of *thread-level parallelism* (TLP) by running the separate threads in parallel, in addition to exploiting ILP among individual instructions within each thread. Unfortunately, different types of applications written to target "conventional" Von Neumann uniprocessors respond to these efforts with varying degrees of success.

1.2   THE APPLICATION PARALLELISM LANDSCAPE

To better understand the potential of CMPs, we survey the parallelism in applications. Figure 1.4 shows a graph of the landscape of parallelism that exists in some typical applications. The *X*-axis shows the various conceptual levels of program parallelism, while the *Y*-axis shows the *granularity* of parallelism, which is the average size of each parallel block of machine instructions between communication and/or synchronization points. The graph shows that as the conceptual level of parallelism rises, the granularity of parallelism also tends to increase although there is a significant overlap in granularity between the different levels.

- *Instruction*: All applications possess some parallelism among individual instructions in the application. This level is not illustrated in the figure, since its granularity is simply single instructions. As was discussed previously, superscalar architectures can take advantage of this type of parallelism.

- *Basic Block*: Small groups of instructions terminated by a branch are known as basic blocks. Traditional architectures have not been able to exploit these usefully to extract any parallelism other than by using ILP extraction among instructions within these small blocks. Effective branch prediction has allowed ILP extraction to be applied

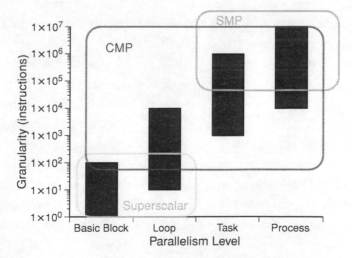

FIGURE 1.4: A summary of the various "ranges" of parallelism that different processor architectures may attempt to exploit.

across a few basic blocks at once, however, greatly increasing the potential for superscalar architectures to find potentially parallel instructions from several basic blocks simultaneously.

- *Loop Iterations*: Each iteration of a typical loop often works with independent data elements, and is therefore an independent chunk of parallel work. (This obviously does not apply to loops with highly dependent loop iterations, such as ones doing pointer-chasing.) On conventional systems, the only way to take advantage of this kind of parallelism is to have a superscalar processor with an instruction window large enough to find parallelism among the individual instructions in multiple loop iterations simultaneously, or a compiler smart enough to interleave instructions from different loop iterations together through an optimization known as *software pipelining*, since hardware cannot parallelize loops directly. Using software tools such as OpenMP, programmers have only had limited success extracting TLP at this level because the loops must be extremely parallel to be divisible into sufficiently large chunks of independent code.

- *Tasks*: Large, independent functions extracted from a single application are known as *tasks*. For example, word processors today often have background tasks to perform spell checking as you type, and web servers typically allocate each page request coming in from the network to its own task. Unlike the previous types of parallelism, only large-scale symmetric multiprocessor (SMP) architectures composed of multiple microprocessor chips have really been able to exploit this level of parallelism, by having programmers manually divide their code into threads that can explicitly exploit TLP using software mechanisms such as POSIX threads (pthreads), since the parallelism is at far too large a scale for superscalar processors to exploit at the ILP level.

- *Processes*: Beyond tasks are completely independent OS processes, all from different applications and each with their own separate virtual address space. Exploiting parallelism at this level is much like exploiting parallelism among tasks, except that the granularity is even larger.

The measure of application performance at the basic block and loop level is usually defined in terms of the *latency* of each task, while at the higher task or process levels performance is usually measured using the *throughput* across multiple tasks or applications, since usually programmers are more interested in the number of tasks completed per unit time than the amount of time allotted to each task.

The advent of CMPs changes the application parallelism landscape. Unlike conventional uniprocessors, multicore chips can use TLP, and can therefore also take advantage of threads to utilize parallelism from the traditional large-grain task and process level parallelism province of SMPs. In addition, due to the much lower communication latencies between processor cores and their ability to incorporate new features that take advantage of these short latencies, such as speculative thread mechanisms, CMPs can attack fine-grained parallelism of loops, tasks and even basic blocks.

1.3   A SIMPLE EXAMPLE: SUPERSCALAR VS. CMP

A good way to illustrate the inherent advantages and disadvantages of a CMP is by comparing performance results obtained for a superscalar uniprocessor with results from a roughly equivalent CMP. Of course, choosing a pair of "roughly equivalent" chips when the underlying architectures are so different can involve some subjective judgment. One way to define "equivalence" is to design the two different chips so that they are about the same size (in terms of silicon area occupied) in an identical process technology, have access to the same off-chip I/O resources, and run at the same clock speed. With models of these two architectures, we can then simulate code execution to see how the performance results differ across a variety of application classes.

To give one example, we can build a CMP made up of four small, simpler processors scaled so that they should occupy about the same amount of area as a single large, superscalar processor. Table 1.1 shows the key characteristics that these two "equivalent" architectures would have if they were actually built. The large superscalar microarchitecture (SS) is essentially a 6-way superscalar version of the MIPS R10,000 processor [4], a prominent processor design from the 1990s that contained features seen in virtually all other leading-edge, out-of-order issue processors designed since. This is fairly comparable to one of today's leading-edge, out-of-order superscalar processors. The chip multiprocessor microarchitecture (CMP), by contrast, is a 4-way CMP composed of four identical 2-way superscalar processors similar to early, in-order superscalar processors from around 1992, such as the original Intel Pentium and DEC Alpha. In both architectures, we model the integer and floating point functional unit result and repeat latencies to be the same as those in the real R10,000.

Since the late 1990s, leading-edge superscalar processors have generally been able to issue about 4–6 instructions per cycle, so our "generic" 6-way issue superscalar architecture is very representative of, or perhaps even slightly more aggressive than, today's most important desktop and server processors such as the Intel Core and AMD Opteron series processors. As the floorplan in Fig. 1.5 indicates, the logic necessary for out-of-order instruction issue and scheduling would dominate the area of the chip, due to the quadratic area impact of supporting very wide instruction issue. Altogether, the logic necessary to handle out-of-order instruction issue occupies about 30% of the die for a processor at this level of complexity. The on-chip memory hierarchy is similar to that used by almost all uniprocessor designs—a small,

TABLE 1.1: Key characteristics of approximately equal-area 6-way superscalar and 4×2-way CMP processor chips.

	6-WAY SS	4×2-WAY MP
♯ of CPUs	1	4
Degree superscalar	6	4×2
♯ of architectural registers	32int/32fp	$4 \times$ 32int/32fp
♯ of physical registers	160int/160fp	$4 \times$ 40int/40fp
♯ of integer functional units	3	4×1
♯ of floating pt. functional units	3	4×1
♯ of load/store ports	8 (one per bank)	4×1
BTB size	2048 entries	4×512 entries
Return stack size	32 entries	4×8 entries
Instruction issue queue size	128 entries	4×8 entries
I cache	32 KB, 2-way S. A.	4×8 KB, 2-way S. A.
D cache	32 KB, 2-way S. A.	4×8 KB, 2-way S. A.
L1 hit time	2 cycles	1 cycle
L1 cache interleaving	8 banks	N/A
Unified L2 cache	256 KB, 2-way S. A.	256 KB, 2-way S. A.
L2 hit time/L1 penalty	4 cycles	5 cycles
Memory latency/L2 penalty	50 cycles	50 cycles

fast level one (L1) cache backed up by a large on-chip level two (L2) cache. The wide issue width requires the L1 cache to support wide instruction fetches from the instruction cache and multiple loads from the data cache during each cycle (using eight independent banks, in this case). The additional overhead of the bank control logic and a crossbar required to arbitrate between the multiple requests sharing the 8 data cache banks adds a cycle to the latency of the L1 cache and increases its area/bit cost. In this example, backing up the 32 KB L1 caches is a

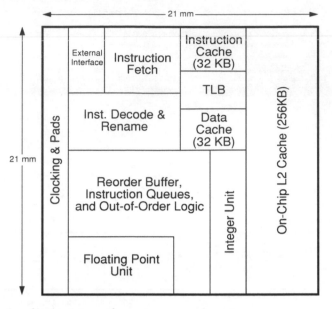

FIGURE 1.5: Floorplan for the 6-issue dynamic superscalar microprocessor.

unified 256 KB L2 cache that takes four cycles to access, which is smaller but faster than most L2 caches in typical processors from industry today.

In contrast, the CMP architecture is made up of four 2-way superscalar processors interconnected by a crossbar that allows the processors to share the L2 cache. On the die, the four processors are arranged in a grid with the L2 cache at one end, as shown in Fig. 1.6. The number of execution units actually increases in the CMP scenario because the 6-way processor included three units of each type, while the 4-way CMP must have four—one for each core. On the other hand, the issue logic becomes *dramatically* smaller, due to the decrease in the number of instruction buffer ports and the smaller number of entries in each instruction buffer. The scaling factors of these two units balance each other out, leaving the entire 4 × 2-way CMP very close to one-fourth of the size of the 6-way processor. More critically, the on-chip cache hierarchy of the multiprocessor is significantly different from the cache hierarchy of the 6-way superscalar processor. Each of the four cores has its own single-banked and single-ported instruction and data caches, and each cache is scaled down by a factor of 4, to 8 KB. Since each cache can only be accessed by a single processor's single load/store unit, no additional overhead is incurred to handle arbitration among independent memory access units. However, since the four processors now share a single L2 cache, that cache requires extra latency to allow time for interprocessor arbitration and crossbar delay.

These two microarchitectures were compared using nine small benchmarks that cover a wide range of possible application classes. Table 1.2 shows the applications: two SPEC95

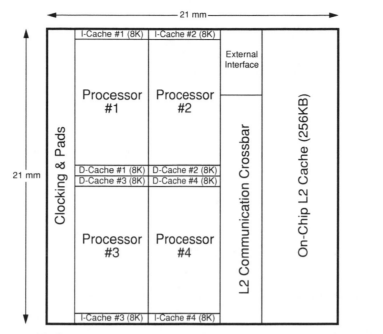

FIGURE 1.6: Floorplan for the 4-way CMP.

TABLE 1.2: Benchmarks used to compare the two equal-area architectures

INTEGER APPLICATIONS	
compress	compresses and uncompresses file in memory
eqntott	translates logic equations into truth tables
m88ksim	Motorola 88000 CPU simulator
MPsim	VCS compiled Verilog simulation of a multiprocessor
FLOATING POINT APPLICATIONS	
applu	solver for parabolic/elliptic partial differential equations
apsi	solves problems of temperature, wind, velocity, and distribution of pollutants
swim	shallow water model with $1K \times 1K$ grid
tomcatv	mesh-generation with Thompson solver
MULTIPROGRAMMING APPLICATION	
pmake	parallel make of gnuchess using C compiler

integer benchmarks (*compress*, *m88ksim*), one SPEC92 integer benchmark (*eqntott*), one other integer application (*MPsim*), four SPEC95 floating point benchmarks (*applu*, *apsi*, *swim*, *tomcatv*), and a multiprogramming application (*pmake*). The applications were parallelized in different ways to run on the CMP microarchitecture. *Compress* was run unmodified on both the SS and CMP microarchitectures, using only one processor of the CMP architecture. *Eqntott* was parallelized manually by modifying a single bit vector comparison routine that is responsible for 90% of the execution time of the application [5]. The CPU simulator (*m88ksim*) was also parallelized manually into three "pipeline stage" threads. Each of the three threads executes a different phase of simulating a different instruction at the same time. This style of parallelization is very similar to the overlap of instruction execution that occurs in hardware pipelining. The *MPsim* application was a Verilog model of a bus-based multiprocessor running under a multithreaded compiled code simulator (*Chronologic VCS-MT*). The multiple threads were specified manually by assigning parts of the model hierarchy to different threads. The *MPsim* application used four closely coupled threads, one for each of the processors in the model. The parallel versions of the SPEC95 floating point benchmarks were automatically generated by the SUIF compiler system [6]. The *pmake* application was a program development workload that consisted of the compile phase of the Modified Andrew Benchmark [7]. The same *pmake* application was executed on both microarchitectures; however, the OS was able to take advantage of the extra processors in the MP microarchitecture to run multiple compilations in parallel.

In order to simplify and speed simulation on this simple demonstration, these are older benchmarks with relatively small data sets compared with those that would be used on real machines today, such as the SPEC CPU 2006 suite [8]. To compensate, the caches and rest of the memory hierarchy were scaled accordingly. For example, while a processor today would tend to have an L2 cache of a few megabytes in size, our simulated systems have L2 caches of only 256 KB. Given this memory system scaling and the fact that we are primarily interested in the performance effects of the processor core architecture differences, the results from this scaled-down simulation example should give a reasonable approximation of what would happen on real systems with larger applications and better memory systems.

1.3.1   Simulation Results

Table 1.3 shows the IPC, branch prediction rates and cache miss rates for one processor of the CMP, while Table 1.4 shows the instructions per cycle (IPC), branch prediction rates, and cache miss rates for the SS microarchitecture. Averages are also presented (geometric for IPC, arithmetic for the others). The cache miss rates are presented in the tables in terms of *misses per completed instruction* (MPCI), including instructions that complete in kernel and user mode. A key observation to note is that when the issue width is increased from 2 to 6 the

TABLE 1.3: Performance of a single 2-issue superscalar processor, with performance similar to an Intel Pentium from the early to mid-1990's

PROGRAM	IPC	BP RATE %	I-CACHE % MPCI	D-CACHE % MPCI	L2 CACHE % MPCI
compress	0.9	85.9	0.0	3.5	1.0
eqntott	1.3	79.8	0.0	0.8	0.7
m88ksim	1.4	91.7	2.2	0.4	0.0
MPsim	0.8	78.7	5.1	2.3	2.3
applu	0.9	79.2	0.0	2.0	1.7
apsi	0.6	95.1	1.0	4.1	2.1
swim	0.9	99.7	0.0	1.2	1.2
tomcatv	0.8	99.6	0.0	7.7	2.2
pmake	1.0	86.2	2.3	2.1	0.4
Average	0.9	88.4	1.2	2.7	1.3

actual IPC increases *by less than a factor of 1.6* for all of the integer and multiprogramming applications. For the floating point applications, more ILP is generally available, and so the IPC varies from a factor of 1.6 for *tomcatv* to 2.4 for *swim*. You should note that, given our assumption of equal—but not necessarily specified—clock cycle times, IPC is directly proportional to the overall performance of the system and is therefore the performance metric of choice.

One of the major causes of processor stalls in a superscalar processor is cache misses. However, cache misses in a dynamically scheduled superscalar processor with speculative execution and nonblocking caches are not straightforward to characterize. The cache misses that occur in a single-issue in-order processor are not necessarily the same as the misses that will occur in the speculative out-of-order processor. In speculative processors there are misses that are caused by speculative instructions that never complete. With nonblocking caches, misses may also occur to lines which already have outstanding misses. Both types of misses tend to inflate the cache miss rate of a speculative out-of-order processor. The second type of miss is mainly responsible for the higher L2 cache miss rates of the 6-issue processor compared to the 2-issue processor, even though the cache sizes are equal.

TABLE 1.4: Performance of the 6-way superscalar processor, which achieves per-cycle performance similar to today's Intel Core or Core 2 out-of-order microprocessors, close to the limit of available and exploitable ILP in many of these programs

PROGRAM	IPC	BP RATE %	I-CACHE % MPCI	D-CACHE % MPCI	L2 CACHE % MPCI
compress	1.2	86.4	0.0	3.9	1.1
eqntott	1.8	80.0	0.0	1.1	1.1
m88ksim	2.3	92.6	0.1	0.0	0.0
MPsim	1.2	81.6	3.4	1.7	2.3
applu	1.7	79.7	0.0	2.8	2.8
apsi	1.2	95.6	0.2	3.1	2.6
swim	2.2	99.8	0.0	2.3	2.5
tomcatv	1.3	99.7	0.0	4.2	4.3
pmake	1.4	82.7	0.7	1.0	0.6
Average	1.5	88.7	0.5	2.2	1.9

Figure 1.7 shows the IPC breakdown for one processor of the CMP with an ideal IPC of two. In addition to the actual IPC achieved, losses in IPC due to data and instruction cache stalls and pipeline stalls are shown. A large percentage of the IPC loss is due to data cache stall time. This is caused by the small size of the primary data cache. *M88ksim*, *MPsim*, and *pmake* have significant instruction cache stall time, which is due to the large instruction working set size of these applications in relation to these small L1 caches. *Pmake* also has multiple processes and significant kernel execution time, which further increases the instruction cache miss rate.

Figure 1.8 shows the IPC breakdown for the SS microarchitecture. A significant amount of IPC is lost due to pipeline stalls. The increase in pipeline stalls relative to the 2-issue processor is due to limited ILP in the applications and the longer L1 data cache hit time. The larger instruction cache in the SS microarchitecture eliminates most of the stalls due to instruction misses for all of these scaled-down applications except *MPsim* and *pmake*. Although the SPEC95 floating point applications have a significant amount of ILP, their performance is limited on the SS microarchitecture due to data cache stalls, which consume over one-half of the available IPC.

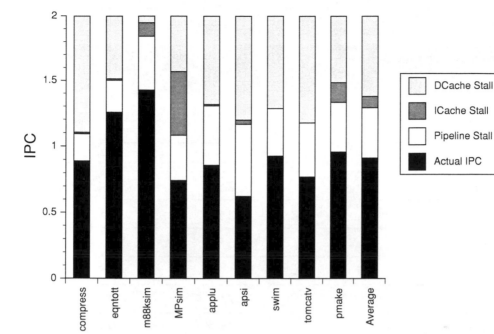

FIGURE 1.7: IPC breakdown for a single 2-issue processor.

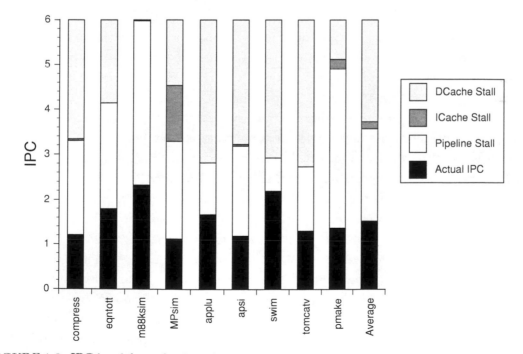

FIGURE 1.8: IPC breakdown for the 6-issue processor.

TABLE 1.5: Performance of the full 4×2-issue CMP

PROGRAM	I-CACHE % MPCI	D-CACHE % MPCI	L2 CACHE % MPCI
compress	0.0	3.5	1.0
eqntott	0.6	5.4	1.2
m88ksim	2.3	3.3	0.0
MPsim	4.8	2.5	3.4
applu	0.0	2.1	1.8
apsi	2.7	6.9	2.0
swim	0.0	1.2	1.5
tomcatv	0.0	7.8	2.5
pmake	2.4	4.6	0.7
Average	1.4	4.1	1.6

Table 1.5 shows cache miss rates for the CMP microarchitecture given in MPCI. To reduce miss-rate effects caused by the idle loop and spinning due to synchronization, the number of completed instructions are those of the original uniprocessor code executing on one processor. Comparing Tables 1.3 and 1.5 shows that for *eqntott*, *m88ksim*, and *apsi* the MP microarchitecture has significantly higher data cache miss rates than the single 2-issue processor. This is due primarily to the high degree of communication present in these applications. Although *pmake* also exhibits an increase in the data cache miss rate, it is caused by process migration from processor to processor in the MP microarchitecture.

Finally, Fig. 1.9 shows the overall performance comparison between the SS and CMP microarchitectures. The performance is measured as the speedup of each microarchitecture relative to a single 2-issue processor running alone. On *compress*, an application with little parallelism, the CMP is able to achieve 75% of the SS performance, even though three of the four processors are idle. For applications with fine-grained parallelism and high communication, such as *eqntott*, *m88ksim*, and *apsi*, performance results on the CMP and SS are similar. Both architectures are able to exploit fine-grained parallelism, although in different ways. The SS microarchitecture relies on the dynamic extraction of ILP from a single thread of control, while the CMP takes advantage of a combination of some ILP and some fine-grained TLP. Both

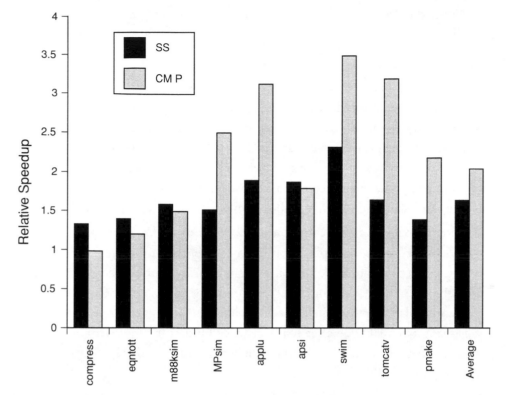

FIGURE 1.9: Performance comparison of SS and CMP.

of these approaches provide a 30–100% performance boost over a single 2-issue processor. However, it should be noted that this boost was completely "free" in the SS architecture, but required some programmer effort to extract TLP in the CMP case. Finally, applications with large amounts of parallelism allow the CMP microarchitecture to take advantage of coarse-grained parallelism and TLP, while the SS can only exploit ILP. For these applications, the CMP is able to significantly outperform the SS microarchitecture, whose ability to dynamically extract parallelism is limited by its single instruction window.

1.4   THIS BOOK: BEYOND BASIC CMPs

Now that we have made the case for why CMP architectures will predominate in the future, in the remainder of this book we examine several techniques that can be used to improve these multicore architectures beyond simple CMP designs that just glue discrete processors together and put them onto single pieces of silicon. The tight, on-chip coupling among processors in CMPs means that several new techniques can be used in multicore designs to improve both applications that are throughput-sensitive, such as most server applications, and ones that are latency-sensitive, such as typical desktop applications. These techniques are organized by the types of applications that they help support:

- In Chapter 2, we look at how CMPs can be designed to effectively address throughput-bound applications, such as our *pmake* example or commercial server workloads, where applications are already heavily threaded and therefore there is significant quantity of natural coarse-grained TLP for the various cores in the CMP to exploit. For these applications, the primary issues are quantitative: one must balance core size, numbers of cores, the amount of on-chip cache, and interconnect bandwidths between all of the on-chip components and to the outside world. Poor balance among these factors can have a significant impact on overall system performance for these applications.

- Chapter 3 focuses on the complimentary problem of designing a CMP for latency-bound general-purpose applications, such as *compress* in the previous example, which require extraction of fine-grained parallelism from sequential code. This chapter focuses on techniques for accelerating legacy uniprocessor code, more or less automatically, by extracting TLP from nominally sequential code. Given the large amount of latency-sensitive sequential code already in existence (and still being written!) and the finite number of engineers familiar with parallel programming, this is a key concern for fully utilizing the CMPs of the future.

- While automated parallelism extraction is helpful for running existing latency-bound applications on CMPs, real programmers will almost always be able to achieve better performance than any automated system, in part because they can adjust algorithms to expose additional parallelism, when necessary. However, conventional parallel programming models, which were originally designed to be implementable with the technology available on multichip multiprocessors of the past, have generally been painful to use. As a result, most programmers have avoided them. Architecture enhancements now possible with CMPs, such as the *transactional memory* described in Chapter 4, can help to simplify parallel programming and make it truly feasible for a much wider range of programmers.

Finally, in Chapter 5 we conclude with a few predictions about the future of CMP architecture design—or, more accurately, the future of all general-purpose microprocessor design, since it is now apparent that we are living in a multicore world.

REFERENCES

[1]   G. E. Moore, "Cramming more components onto integrated circuits," *Electronics*, vol. 38, no. 8, pp. 114–117, Apr. 19, 1965.

[2] J. L. Hennessy and D. A. Patterson, *Computer Architecture: A Quantitative Approach*, 3rd edition. San Francisco, CA: Morgan Kaufmann, 2003.

[3] D. W. Wall, "Limits of instruction-level parallelism," WRL Research Report 93/6, Digital Western Research Laboratory, Palo Alto, CA, 1993.

[4] K. Yeager et al., "R10 000 superscalar microprocessor," presented at *Hot Chips VII*, Stanford, CA, 1995.

[5] B. A. Nayfeh, L. Hammond, and K. Olukotun, "Evaluating alternatives for a multiprocessor microprocessor," in *Proceedings of 23rd Int. Symp. Computer Architecture*, Philadelphia, PA, 1996, pp. 66–77.

[6] S. Amarasinghe et al., "Hot compilers for future hot chips," in *Hot Chips VII*, Stanford, CA, Aug. 1995. http://www.hotchips.org/archives/

[7] J. Ousterhout, "Why aren't operating systems getting faster as fast as hardware?" in *Summer 1990 USENIX Conference*, June 1990, pp. 247–256.

[8] Standard Performance Evaluation Corporation, SPEC, http://www.spec.org, Warrenton, VA.

CHAPTER 2

Improving Throughput

With the rise of the Internet, demand has dramatically increased for servers capable of handling a multitude of independent requests arriving rapidly over the network. Since individual network requests are typically completely independent tasks, whether those requests are for web pages, database access, or file service, they are typically spread across many separate computers built using high-performance conventional microprocessors, a technique that has been used at places like Google [1] for years, in order to match the overall computation *throughput* to the input request rate. As the number of requests increased over time, more servers were added to the collection. In addition to adding more servers, it has also been possible to replace some or all of the separate servers with multiprocessors. Most existing multiprocessors consist of two or more separate processors connected using a common bus, switch hub, or network to shared memory and I/O devices. The overall multiprocessor system can usually be physically smaller and use less power than an equivalent set of uniprocessor systems because physically large components such as memory, hard drives, and power supplies can be shared by some or all of the processors.

Pressure has increased over time to achieve more performance per unit volume of data center space and per Watt, since data centers have finite room for servers and their electric bills can be staggering. In response, the server manufacturers have tried to save space by adopting denser server packaging solutions, such as blade servers, and by switching to multiprocessors that can share components. In addition to saving space, server manufacturers have reduced power consumption through the sharing of power-hungry memory and I/O components. However, these short-term solutions are reaching their practical limits as systems are reaching the maximum component density that can still be effectively air cooled, a sort of "power wall" imposed by the laws of physics. As a result, the next stage of development for these systems involves moving away from packing larger numbers of single processor chips into boxes and switching to chip multiprocessors, instead [2].

The first CMPs targeted toward the server market implement two or more conventional superscalar processors together on a single die [3–6]. The primary motivation for this is reduced volume—now multiple processors can fit in the space where formerly only one could, so overall performance per unit volume can be increased. There is also some savings in power because all of the processors on a single die can share a single connection to the rest of the system, reducing the amount of high-speed communication infrastructure required, in addition to the sharing possible with a conventional multiprocessor.

Further savings in power can be achieved by taking advantage of the fact that while server workloads require high throughput, the *latency* of each request is generally not as critical [7]. Most users will not be bothered if their web pages take a fraction of a second longer to load, but they will complain if the website drops page requests because it does not have enough throughput capacity. A CMP-based system can be designed to take advantage of this situation. When a 2-way CMP replaces a uniprocessor in a system, it is possible to achieve essentially the same or better throughput on server-oriented workloads with just *half* of the original clock speed. Each server request will take up to twice as long to process due to the reduced clock rate, although with many server applications the slowdown will be much less than this, because request processing time is more often limited by memory or disk performance, which remains the same in both systems, instead of processor performance. However, since two requests can now be processed simultaneously, the overall throughput will now be the same or better. Overall, even if the performance is the same or only a little better, this adjustment is still very advantageous at the system level. The lower clock rate allows the design of a system with a significantly lower power supply voltage, often a nearly linear reduction. Since power is directly proportional to frequency and proportional to the *square* of the voltage, however, the power required to obtain the original performance is much lower—potentially as low as a quarter (half of the power due to the frequency reduction and one-half squared or a quarter of the power due to the voltage, for a total of one-eighth of the power per processor, so the power required for both processors together is one-quarter), although the potential savings will usually not quite achieve this level due to the limits of static power dissipation and any minimum voltage levels required by the underlying transistors.

In a similar manner, CMP designers can reduce the *complexity* of the individual cores, in addition to their clock speed. Instead of a couple of large and complex processor cores, CMP designers can just as easily—if not *more* easily—build a CMP using many more simple cores, instead, in the same area. Why? Commercial server applications exhibit high cache miss rates, large memory footprints, and low instruction level parallelism (ILP), which leads to poor utilization on traditionally ILP-focused processors [9]. Processor performance increases in the recent past have come predominately from optimizations that burn large amounts of power for relatively little performance gain in the presence of low ILP applications. The current 3-to-6-way out-of-order superscalar server processors perform massive amounts of speculative execution, with this speculation translating into fairly modest performance gains on server workloads [8]. Simpler cores running these same applications will undoubtedly take longer to perform the same task, but the relatively small reduction in latency performance is rewarded

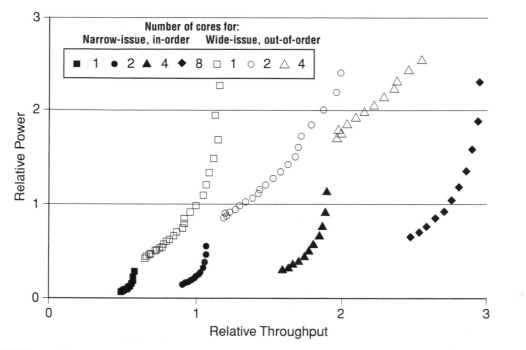

FIGURE 2.1: Comparison of power usage by equivalent narrow-issue, in-order and wide-issue, out-of-order processors on throughput-oriented software, from [10]. Each type of processor is tested with different numbers of cores and across wide ranges of supply voltage and clock frequency.

not only with higher overall throughput (due to the fact that more cores can be implemented on each processor chip) but also with better performance/Watt (due to the elimination of power-wasting speculative execution).

Figure 2.1 shows a comparison between simple, in-order and complex, out-of-order processors across a wide range of voltages and clock rates, as described in [10]. For any given desired level of throughput (*x*-axis), CMPs made from simple, in-order cores are able to achieve the same performance using only a fraction of the power of an equivalent machine made using complex, out-of-order processors due to their significantly better performance/Watt figures. For example, at the reference "throughput = 1" point, the in-order processor only requires about 20% of the power needed by the out-of-order processor to achieve the same performance, at the expense of requiring two parallel software threads instead of just one. Alternatively, looking at the same data from a position of fixed power limits, the 8-thread in-order and 2-thread out-of-order processors require similar amounts of power across the voltage range studied, but the 8-thread system is 1.5–2.25 times faster than the 2-thread system across the range. The rest of this chapter investigates how this simple-core performance/Watt advantage works in practice.

2.1 SIMPLE CORES AND SERVER APPLICATIONS

In general, the simpler the pipeline, the lower the power. This is because deeper and more complex pipelines require more transistors, switching at a higher frequency, to complete the same function as a shallower, simple pipeline. In addition to the simple linear increase in the number of the pipeline registers associated with adding pipeline stages, the cycle time overhead from issues such as pipeline stage work imbalance, additional register setup/hold times for each stage of registers, and the higher potential for clock skew at high frequencies causes a superlinear increase in the number of registers required in any real system simply due to overhead. These additional overheads are fixed regardless of cycle time, so that as the processor becomes more deeply pipelined and the cycle time decreases, these overheads become a larger and larger portion of the clock cycle and make additional pipelining even more expensive in terms of the number of transistors (and therefore power) required [11]. A deeper and more complex pipeline also requires that more of the instructions be executed speculatively, and when that speculation does not pay off, the extra power consumed is wasted. Furthermore, additional power is required to get the processor back on the right execution path. As a consequence, shallow, simple pipelines inherently have a power advantage over today's deep, complex pipelines. One can take this argument to its extreme, arguing that the most power-efficient processor is one that is not pipelined at all. However, the flaw in taking the argument to its extreme is that all the power in a processor is not dynamic power generated by switching the state of the individual transistors. Leakage power is also expended simply keeping the transistors in their current state. Historically, this static power resulting from transistor leakage has been a tiny fraction of the total processor power budget. However, scaling transistor geometries results in increasing leakage power, making static power a significant component of total power in modern processors. The presence of static power moves the most power-efficient point for a processor from one that is not pipelined to one that has a shallow or modest-depth pipeline.

2.1.1 The Need for Multithreading within Processors

With most server applications, there is an abundant amount of application thread-level parallelism [9], but low instruction-level parallelism and high cache miss rates. Due to the high cache miss rates and low instruction-level parallelism, most modern processors are idle a significant portion of the time while running these applications. Clock gating can be used to reduce the active power of the idle blocks; however, the static and clock distribution power remains. Therefore, using *multithreading* to keep otherwise idle on-chip resources busy results in a significant performance/Watt improvement by boosting performance by a large amount while increasing active power by a smaller amount. Multithreading is simply the process of adding hardware to each processor to allow it to execute instructions from multiple threads, either one at a time or simultaneously, without requiring OS or other software intervention to perform each thread switch—the conventional method of handling threading. Adding multithreading to a processor is not free; processor die area is increased by the replication of register files and other architecturally visible state (such as trap stack registers, MMU registers, etc.) and the addition of logic to the pipeline to switch between threads. However, this additional area is fairly modest. Adding an additional thread to a single-threaded processor costs 4–7% in terms of area [12–14], and each additional thread adds a similar percentage increase in area.

One question facing architects is what style of multithreading to employ. There are three major techniques: *coarse-grained* [15], where a processor runs from a single thread until a long-latency stall such as a cache miss triggers a thread switch; *fine-grained* (or interleaved [16]), where a processor switches between several "active" (not stalled) threads every cycle; and *simultaneous multithreading* [17], where instructions may be issued from multiple threads during the same cycle within a superscalar processor core. Note that for a single-issue processor, fine-grained and simultaneous multithreading are equivalent. Coarse-grained multithreading has the drawback that short-latency events (such as pipeline hazards or shared execution resource conflicts) cannot be hidden simply by switching threads due to the multicycle cost of switching between threads [16], and these short-latency events are a significant contributor to CPI in real commercial systems, where applications are plagued by many brief stalls caused by difficult-to-predict branches and loads followed soon thereafter by dependent instructions. Multicycle primary caches, used in some modern processors, can cause these load-use combinations to almost always stall for a cycle or two, and even secondary cache hits can be handled quickly enough on many systems (10–20 cycles) to make hiding them with coarse-grained multithreading impractical. As a result, most multithreaded processors to date have employed fine-grained [11] or simultaneous multithreading [14, 18], and it is likely that future processors will as well.

2.1.2   Maximizing the Number of Cores on the Die

Once each processor core has sufficient threads to keep the pipeline utilized, multiple processor cores can be added to the chip to increase the total thread count. However, the number of cores that can be added to the chip is strongly dependent on the size of each individual core. As was discussed in the first chapter, there are two main options for the processor cores: a complex, heavyweight processor core, which emphasizes low thread latency over core area, and a simple, lightweight processor core, which emphasizes core area over thread latency. Thus, a CMP can either employ a smaller number of complex cores, emphasizing individual thread completion time but sacrificing aggregate thread throughput, or employ a larger number of simple processor cores, emphasizing aggregate thread throughput but degrading individual thread completion time. The simple core approach has intuitive appeal, as a simple, scalar processor can be built in much less area than a complex, superscalar processor, and still provide similar sustained performance on large commercial applications, which are mostly memory-latency bound. In addition, the performance/Watt of the simple core processor will be higher than that of the superscalar processor, which relies on a large level of speculation resources to achieve high rates of instruction-level parallelism.

2.1.3   Providing Sufficient Cache and Memory Bandwidth

The final component that is required for improved throughput and performance/Watt from a server CMP is sufficient cache and memory bandwidth. When multithreading is added to a processor without providing sufficient bandwidth for the memory demands created by the increased number of threads, only modest gains in performance (and sometimes even slowdowns) will be seen [14].

2.2   CASE STUDIES OF THROUGHPUT-ORIENTED CMPs

To show how these rules work in reality, three case studies of CMPs designed from the ground up for executing throughput-based workloads well are used: the Piranha project and the Sun's Niagara (UltraSPARC T1) and Niagara 2 (UltraSPARC T2) processors.

2.2.1   Example 1: The Piranha Server CMP

The Piranha design [7] was one of the first to adopt the model of a CMP composed of several simpler processor cores, in 2000. Figure 2.2 shows the block diagram of a single Piranha processing chip. Each Alpha CPU core (CPU) is directly connected to dedicated instruction (iL1) and data cache (dL1) modules. These first-level caches interface to other modules through

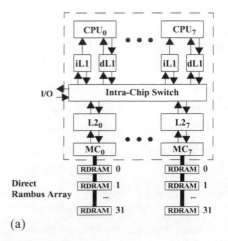

(a)

FIGURE 2.2: Block diagram of a CMP Piranha processing node.

the intrachip switch (ICS). On the other side of the ICS is a logically shared second-level cache (L2) that is interleaved into eight separate modules, each with its own controller, on-chip tag, and data storage. Attached to each L2 module is a memory controller (MC), which directly interfaces to one bank of up to 32 direct Rambus DRAM chips. Each memory bank provides a bandwidth of 1.6 GB/s, leading to an aggregate bandwidth of 12.8 GB/s. The Piranha design also includes an on-chip interconnect controller supporting shared memory across multiple Piranha chips, but this discussion will focus on a single Piranha chip.

2.2.1.1 Processor Core

The processor core uses a single-issue, in-order design capable of executing the Alpha instruction set [19]. It consists of a 500 MHz pipelined datapath with hardware support for floating-point operations. The pipeline has eight stages: instruction fetch, register-read, ALU 1 through 5, and write-back. The 5-stage ALU supports pipelined floating-point and multiply instructions. However, most instructions execute in a single cycle. The processor core includes several performance enhancing features including a branch target buffer, precompute logic for branch conditions, and a fully bypassed datapath. The processor core interfaces to separate first-level instruction and data caches designed for single-cycle latency. The first-level caches are both 64 KB 2-way set-associative, blocking caches with virtual indices and physical tags. The L1 cache modules include tag compare logic, instruction and data TLBs (256 entries, 4-way associative), and a store buffer (data cache only). A 2-bit state field is maintained per cache line, corresponding to the four states in a typical MESI protocol. Both the instruction and data caches are kept coherent by hardware.

Processor cores are connected to each other and lower levels of memory using the ICS, a large crossbar. The ICS is also the primary facility for decomposing the Piranha design into relatively independent, isolated modules. The transactional nature of the ICS allowed the Piranha designers to add or remove pipeline stages during the design of various modules without compromising the overall Piranha timing, avoiding the tight timing dependences often found across entire large uniprocessor designs.

2.2.1.2 Second-level Cache

Piranha's second-level cache (L2) is a 1 MB unified instruction/data cache which is physically partitioned into eight banks and is logically shared among all CPUs. The L2 banks are interleaved using the lower address bits of a cache line's physical address (64-byte line). Each bank is 8-way set-associative and uses a round-robin (or least-recently-loaded) replacement policy if an invalid block is not available. Each bank has its own control logic, an interface to its private memory controller, and an ICS interface used to communicate with other chip modules. The L2 controllers are responsible for maintaining intrachip coherence.

Since Piranha's aggregate L1 capacity is 1 MB, maintaining data inclusion in the 1 MB L2 can potentially waste its full capacity with duplicate data. Therefore, Piranha employs a noninclusive L2 cache [20]. To simplify intrachip coherence and avoid the use of snooping at L1 caches, Piranha keeps a duplicate copy of the L1 tags and state at the L2 controllers. Each controller maintains tag/state information for L1 lines that map to it, given the address interleaving. The total overhead for the duplicate L1 tag/state across all controllers is less than 1/32 of the total on-chip memory.

In order to lower miss latency and best utilize the L2 capacity, L1 misses that also miss in the L2 are filled directly from memory without allocating a line in the L2. The L2 effectively behaves as a very large victim cache that is filled only when data is replaced from the L1s. Hence, even clean lines that are replaced from an L1 may cause a write-back to the L2. To avoid unnecessary write-backs when multiple L1s have copies of the same line, the duplicate L1 state is extended to include the notion of ownership. The owner of a line is either the L2 (when it has a valid copy), an L1 in the exclusive state, or one of the L1s (typically the last requester) when there are multiple sharers. Based on this information, the L2 makes the decision of whether an L1 should write-back its data and piggybacks this information with the reply to the L1's request (that caused the replacement). In the case of multiple sharers, a write-back happens only when an owner L1 replaces the data. The above approach provides Piranha with a near-optimal replacement policy without affecting the L2 hit time.

The L2 controllers are responsible for enforcing coherence within a chip. Each controller has complete and exact information about the on-chip cached copies for the subset of lines that map to it. On every L2 access, the duplicate L1 tag/state and the tag/state of the L2 itself are checked in parallel. Therefore, the intrachip coherence has similarities to a full-map centralized directory-based protocol.

A memory request from an L1 is sent to the appropriate L2 bank based on the address interleaving. Depending on the state at the L2, the L2 can possibly service the request directly, forward the request to a local (owner) L1, or obtain the data from memory through the memory controller.

2.2.1.3 Memory Controller

Piranha has a high bandwidth, low latency memory system based on direct Rambus RDRAM. Each L2 bank has one memory controller and associated RDRAM channel, for a total of eight memory controllers. Each Rambus channel can support up to 32 RDRAM chips. In the 64 Mbit memory chip generation, each Piranha processing chip supports a total of 2 GB of physical memory (8 GB/32 GB with 256 Mb/1 Gb chips). Each RDRAM channel has a maximum data rate of 1.6 GB/s, providing a maximum local memory bandwidth of 12.8 GB/s per processing chip. The latency for a random access to memory over the RDRAM channel is 60 ns for the critical word, and an additional 30 ns for the rest of the cache line. Unlike other Piranha chip modules, the memory controller does not have direct access to the intrachip switch. Access to memory is controlled by and routed through the corresponding L2 controller. The L2 controller can issue cache-line-sized read/write requests to the corresponding memory controller. A sophisticated open-page policy is employed to reduce the access latency from 60 ns to 40 ns for page hits.

2.2.1.4 Server Workloads

Piranha was evaluated with two different workloads running on top of an Oracle 7.3.2 commercial database server. The first was an OLTP workload modeled after the TPC-B benchmark [21]. This benchmark models a banking database system that keeps track of customers' account balances, as well as balances per branch and teller. Each transaction updates a randomly chosen account balance, which includes updating the balance of the branch the customer belongs to and the teller from which the transaction is submitted. It also adds an entry to the history table, which keeps a record of all submitted transactions. The DSS workload is modeled after Query 6 of the TPC-D benchmark [22]. The TPC-D benchmark represents the activities of a business that sells a large number of products on a worldwide scale. It consists of several interrelated tables that keep information such as parts and customer orders. Query 6 scans the largest table in the database to assess the increase in revenue that would have resulted if some discounts were eliminated. The behavior of this query is representative of other TPC-D queries [23], though some queries exhibit less parallelism.

The OLTP and DSS workloads were set up and scaled in a way similar to a previous study that validated such scaling [23]. The TPC-B database had 40 branches with a shared-memory segment (SGA) size of approximately 600 MB (the size of the metadata area is about 80 MB), and the runs consisted of 500 transactions after a warm-up period. To utilize the processors usefully while waiting for I/O to complete, eight server processes per processor are allocated to run transactions in parallel. For DSS, Oracle is configured to use the Parallel Query Optimization option, which allows the database engine to decompose the query into multiple subtasks and assign each one to an Oracle server process. The DSS experiments use an in-memory 500 MB database, and the queries are parallelized to generate four server processes per processor.

2.2.1.5 Simulation Environment

Simulations were performed on the SimOS-Alpha environment (the Alpha port of SimOS [24]), which was used in a previous study of commercial applications and has been validated against Alpha multiprocessor hardware [23]. SimOS-Alpha is a full system simulation environment that simulates the hardware components of Alpha-based multiprocessors (processors, MMU, caches, disks, console) in enough detail to run Alpha system software. The ability to simulate both user and system code under SimOS–Alpha is essential given the rich level of system interactions exhibited by commercial workloads. For example, for the OLTP runs in the Piranha study, the kernel component is approximately 25% of the total execution time (user and kernel).

Table 2.1 presents the processor and memory system parameters for the different processor configurations studied. For the out-of-order microprocessor being compared against Piranha, an aggressive design (at that time) similar to Alpha 21364 was selected which integrated a 1 GHz out-of-order core, two levels of caches, memory controller, coherence hardware, and network router all on a single die (with area comparable to Piranha's processing chip). The frequency and L2 cache size in Piranha were limited by their use of an ASIC process, so the Piranha team also explored the possibility of building a full-custom design. Given the simple

TABLE 2.1: Parameters for different processor designs

PARAMETER	PIRANHA (P8)	NEXT-GENERATION MICROPROCESSOR (OOO)	FULL-CUSTOM PIRANHA (P8F)
Processor Speed	500 MHz	1 GHz	1.25 GHz
Type	In-order	Out-of-order	In-order
Issue Width	1	4	1
Instruction Window Size	–	64	–
Cache Line Size	64 bytes	64 bytes	64 bytes
L1 Cache Size	64 KB	64 KB	64 KB
L1 Cache Associativity	2-way	2-way	2-way
L2 Cache Size	1 MB	1.5 MB	1.5 MB
L2 Cache Associativity	8-way	6-way	6-way
L2 Hit/L2 Fwd Latency	16 ns/24 ns	12 ns/ NA	12 ns/16 ns
Local Memory Latency	80 ns	80 ns	80 ns
Remote Memory Latency	120 ns	120 ns	120 ns
Remote Dirty Latency	180 ns	180 ns	180 ns

single-issue in-order pipeline, the Piranha team estimated that a full-custom approach would lead to a 25% faster clock frequency than a 4-issue out-of-order design.

Table 2.1 also shows the memory latencies for the different configurations. Due to the lack of inclusion in Piranha's L2 cache, the table includes two latency parameters corresponding to either the L2 servicing the request (L2 Hit) or the request being forwarded to be serviced by another on-chip L1 (L2 Fwd). As shown in Table 2.1, the Piranha prototype had a higher L2 hit latency than a full-custom processor due to the use of slower ASIC SRAM cells.

2.2.1.6 Performance Evaluation of Piranha

The first evaluation compares Piranha and the aggressive out-of-order processor (OOO in Table 2.1) on the OLTP and DSS database workloads, while the second compares the full-custom Piranha design (P8F in Table 2.1) against the full-custom OOO to more fairly judge the merits of the Piranha architecture.

Figure 2.3 shows the results for single-chip configurations for both OLTP and DSS. The Piranha team studied four configurations here: a hypothetical single-CPU Piranha chip (P1), the Alpha 21364-like out-of-order processor (OOO), a hypothetical single-issue in-order processor otherwise identical to OOO (INO), and the actual eight-CPU Piranha chip (P8). The P1 and INO configurations were used to better isolate the various factors that contribute to the performance differences between OOO and P8. The figure shows execution time normalized to that of OOO. The execution time is divided into CPU busy time, L2 hit stall time, and L2 miss stall time. For the P8 configuration, the L2 hit stall time includes both L2 hits as well as forwarded L2 requests served by an L1 (see L2 Fwd latency in Table 2.1). Focusing on the OLTP results, OOO outperforms P1 by a factor of approximately 2.3. The INO result shows that the faster frequency (1 GHz vs. 500 MHz) and lower L2 hit latency (12 ns in INO/OOO

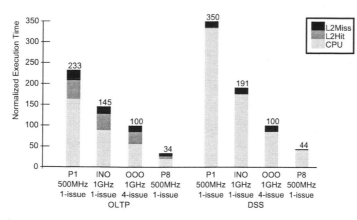

FIGURE 2.3: Estimated performance of a single-chip Piranha (8 CPUs/chip) versus a 1 GHz out-of-order processor.

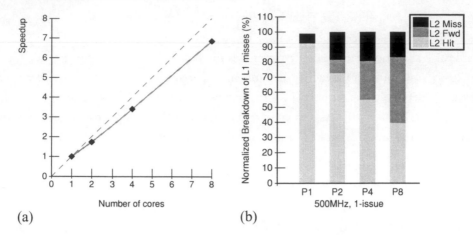

FIGURE 2.4: Piranha's (a) speedup and (b) L1 miss breakdown for OLTP.

vs. 16/24 ns in P1/P8) alone account for an improvement factor of 1.6. The wider-issue and out-of-order features provide the remaining factor of 1.45. Validating the earlier discussion on the merits of designing a CMP from larger numbers of simple cores, the integration of eight of the Piranha CPUs into the single chip Piranha (P8) leads to Piranha outperforming OOO by almost a factor of 3.

As shown in Fig. 2.4(a), the reason for Piranha's exceptional performance on OLTP is that it achieves a speedup of nearly seven with eight on-chip CPUs relative to a single CPU (P1). This speedup arises from the abundance of thread-level parallelism in OLTP, along with the extremely tight-coupling of the on-chip CPUs through the shared second-level cache (leading to small communication latencies), and the effectiveness of the on-chip caches in Piranha. The last effect is clearly observed in Fig. 2.4(b) which shows the behavior of the L2 cache as more on-chip CPUs are added. This figure shows a breakdown of the total number of L1 misses that are served by the L2 (L2 Hit), forwarded to another on-chip L1 (L2 Fwd), or served by the memory (L2 Miss). Although the fraction of L2 hits drops from about 90% to under 40% going from one to eight CPUs, the fraction of L2 misses that go to memory remains constant past a single CPU at under 20%. In fact, adding CPUs (and their corresponding L1s) in Piranha's noninclusive cache hierarchy actually increases the amount of on-chip memory (P8 doubles the on-chip memory compared to P1), which partially offsets the effects of the increased pressure on the L2. The overall trend is that as the number of CPUs increases, more L2 misses are served by other L1s instead of going to memory. Even though "L2 Fwd" accesses are slower than L2 Hits (24 ns vs. 16 ns), they are still much faster than a memory access (80 ns). Overall, Piranha's noninclusion policy is effective in utilizing the total amount of on-chip cache memory (i.e., both L1 and L2) to contain the working set of a parallel application.

In addition to the above on-chip memory effects, the simultaneous execution of multiple threads enables Piranha to tolerate long latency misses by allowing threads in other CPUs to proceed independently. As a result, a Piranha chip can sustain a relatively high CPU utilization level despite having about three times the number of L2 misses as OOO. On-chip and off-chip bandwidths are also not a problem even with eight CPUs because OLTP is primarily latency bound. Finally, OLTP workloads have been shown to exhibit constructive interference in the instruction and data streams [8], and this works to the benefit of Piranha.

Referring back to Fig. 2.3, it is apparent that Piranha (P8) also outperformed OOO for DSS, although by a narrower margin than for OLTP (2.3 times). The main reason for the narrower margin comes from the workload's smaller memory stall component (under 5% of execution time) and better utilization of issue slots in a wide-issue out-of-order processor. DSS is composed of tight loops that exploit spatial locality in the data cache and have a smaller instruction footprint than OLTP. Since most of the execution time in DSS is spent in the CPU, OOO's faster clock speed alone nearly doubles its performance compared to P1 (P1 vs. INO), with almost another doubling due to wider-issue and out-of-order execution (INO vs. OOO). However, the smaller memory stall component of DSS also benefits Piranha, as it achieves near-linear speedup with eight CPUs (P8) over a single CPU (P1).

To more fairly judge the potential of the Piranha approach by separating the Piranha architecture from the prototype implementation, the Piranha team also evaluated the performance of a full-custom implementation (see Table 2.1 for P8F parameters). Figure 2.5 compares the performance of a full-custom Piranha with that of OOO, both in single-chip configurations. The figure shows the faster full-custom implementation further boosted Piranha's performance to 5.0 times over OOO in OLTP and 5.3 times in DSS. DSS saw particularly substantial gains

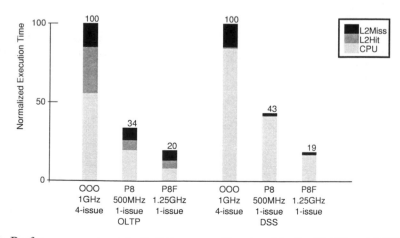

FIGURE 2.5: Performance potential of a full-custom Piranha chip for OLTP and DSS.

since its performance was dominated by CPU busy time, and therefore it benefited more from the 150% boost in clock speed (P8 vs. P8F). The gains in OLTP were also mostly from the faster clock cycle, since the relative improvement in memory latencies compared to processor speed was smaller. The Piranha team concluded their performance study by stating: "Overall the Piranha architecture seems to be a better match for the underlying thread-level parallelism available in database workloads than a typical next generation out-of-order superscalar processor design which relies on its ability to extract instruction-level parallelism." The Piranha design embodied all the rules for CMP design discussed earlier, with the exception of supporting multithreading. For a design that added in multithreading to the mix, we turn to the second example, the Niagara processor from Sun Microsystems.

2.2.2 Example 2: The Niagara Server CMP

The Niagara processor from Sun Microsystems [11], illustrated in Fig. 2.6, is also a good example of a simple core CMP that is designed specifically for high throughput and excellent

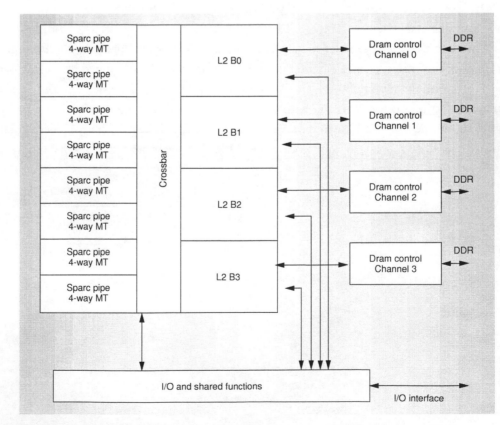

FIGURE 2.6: Niagara-1 block diagram.

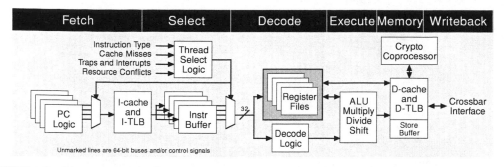

FIGURE 2.7: Niagara core pipeline.

performance/Watt on server workloads. Unlike the Piranha, the Niagara CMP became an actual product (the Sun UltraSPARC T1); it has therefore been investigated in much more detail using real silicon. Like Piranha, Niagara employs eight scalar, shallow pipeline processors on a single die. The pipeline on Niagara is quite shallow, only six stages deep, and employs very little speculation, eschewing even the branch prediction that was present in Piranha. The Niagara pipeline is illustrated in Fig. 2.7.

Besides the obvious differences in instruction set architecture, frequency, and cache sizes, Niagara differs from Piranha by embracing multithreading. Each Niagara processor supports four threads in hardware, resulting in a total of 32 threads on the CPU. The Niagara processor employs fine-grain multithreading, and the processor hides memory and pipeline stalls on a given thread by scheduling the other threads in the group onto the pipeline with the zero-cycle switch penalty characteristic of fine-grain multithreading.

At a high level, the cache and memory subsystem in Niagara is similar to Piranha, with each bank of the level-two cache talking directly to a single memory controller. Niagara has a four-banked L2 cache instead of Piranha's eight-banked L2 cache. A crossbar also interconnects the processor and the L2 cache on Niagara. The memory controllers on Niagara have advanced along with main memory technology from controlling Piranha's Rambus to controlling DDR2 SDRAM. Niagara includes an on-chip IO controller, which provides a Sun-proprietary JBUS I/O interface.

Niagara is built in a 90 nm process from TI, compared to the 180 nm process targeted by Piranha, and as such Niagara is able to include a much larger 3 MB level-two cache and run at a higher frequency of 1.2 GHz. A die photo of Niagara is shown in Fig. 2.8. A key challenge in building a CMP is minimizing the distance between the processor cores and the shared secondary cache. Niagara addresses this issue by placing four cores at the top of the die, four cores at the bottom of the die, with the crossbar and the four L2 tag arrays located in the center of the die. The four L2 data banks and memory I/O pads surround the core, crossbar, and

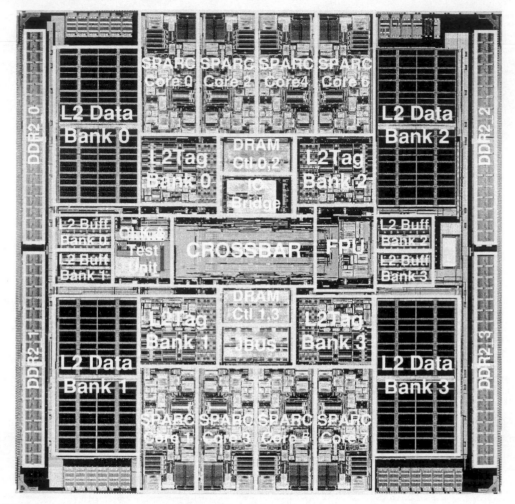

FIGURE 2.8: Niagara-1 die microphotograph.

L2 tags on both the left and right sides of the die. Filling in the space between the major blocks are the four memory (DRAM) controllers, the JBUS controller, internal memory-mapped I/O registers (I/O Bridge), a single-shared floating point (FPU), and clock, test, and data buffering blocks.

2.2.2.1 Multithreading on Niagara

Niagara employs fine-grained multithreading (which is equivalent to simultaneous multithreading for the Niagara scalar pipeline), switching between available threads each cycle, with priority given to the least recently used thread. Threads on Niagara can become unavailable because of long-latency instructions such as loads, branches, multiply, and divide. They also become

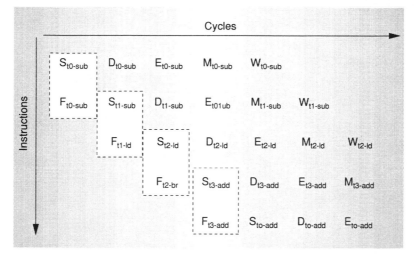

FIGURE 2.9: Thread selection: all threads available.

unavailable because of pipeline "stalls" such as cache misses, traps, and resource conflicts. In Niagara, the thread scheduler assumes that loads are cache hits, and can therefore issue a dependent instruction from the same thread speculatively once the three-cycle load-to-use cost of Niagara has been satisfied. However, such a speculative thread is assigned a lower priority for instruction issue than a thread that can issue a nonspeculative instruction. Along with the fetch of the next instruction pair into an instruction buffer, this speculative issuing of an instruction following a load is the only speculation performed by the Niagara processor.

Figure 2.9 indicates the operation of a Niagara processor when all threads are available. In the figure, you can track the progress of an instruction through the pipeline by reading left-to-right along a row in the diagram. Each row represents a new instruction fetched into the pipe from the instruction cache, sequenced from top to bottom in the figure. The notation St0-sub refers to a Subtract instruction from thread 0 in the S stage of the pipe. In the example, the t0-sub is issued down the pipe. As the other three threads become available, the thread state machine selects thread 1 and deselects thread 0. In the second cycle, similarly, the pipeline executes the t1-sub and selects t2-ld (load instruction from thread 2) for issue in the following cycle. When t3-add is in the S stage, all threads have been executed, and for the next cycle the pipeline selects the least recently used thread, thread 0. When the thread-select stage chooses a thread for execution, the fetch stage chooses the same thread for instruction cache access.

Figure 2.10 indicates the operation when only two threads are available. Here thread 0 and thread 1 are available, while thread 2 and thread 3 are not. The t0-ld in the thread-select stage in the example is a long-latency operation. Therefore it causes the deselection of thread 0. The t0-ld itself, however, issues down the pipe. In the second cycle, since thread 1 is available,

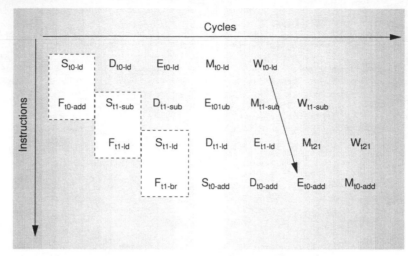

FIGURE 2.10: Thread selection: only two threads available. The ADD instruction from thread 0 is speculatively switched into the pipeline before the hit/miss for the load instruction has been determined.

the thread scheduler switches it in. At this time, there are no other threads available and the t1-sub is a single-cycle operation, so thread 1 continues to be selected for the next cycle. The subsequent instruction is a t1-ld and causes the deselection of thread 1 for the fourth cycle. At this time only thread 0 is speculatively available and therefore can be selected. If the first t0-ld was a hit, data can bypass to the dependent t0-add in the execute stage. If the load missed, the pipeline flushes the subsequent t0-add to the thread select stage instruction buffer, and the instruction reissues when the load returns from the L2 cache.

Multithreading is a powerful architectural tool. On Niagara, adding four threads to the core increased the core area by 19–20% to a total of 16.35 mm^2 in a 90 nm process, when the area for the per-core cryptography unit is excluded. As will be shown later in this chapter, the addition of these four threads may result in a two and half- or even threefold speedup for many large-scale commercial applications. These large performance gains translate into an increase in performance/Watt for the multithreaded processor, as the power does not increase threefold. Instead the core power increase is roughly 1.2 (static and clocking power) + 3.0 (dynamic power), where the 1.2 increase in static and clocking power results from the 20% increase in processor core area for the addition of four threads. Note that the threefold increase in core dynamic power is an upper bound that assumes perfect clock gating and no speculation being performed when a thread is idle. In addition to the core power increase, the multithreaded processor provides an increased load on the level-two (L2) cache and memory. This load can be even larger than the threefold increase in performance, as destructive interference between threads can result in higher L1 and L2 miss rates. For an application experiencing a threefold speedup due to multithreading, the L2 static power remains the same, while the active power could increase by a factor of greater than 3 due to the increase in the L2 load. (In [26], the L2 dynamic power for a database workload was found to increase by 3.6.) Likewise, for memory, the static power (including refresh power) remains the same, while active power could increase by a factor greater than 3. (In [26], memory power for the database workload increased by a factor of 5.1.) However, for both the L2 and memory, the static power is a large component of the total power, and thus even within the cache and memory subsystems, a threefold increase in performance from multithreading will likely exceed the increase in power, resulting in an increase in performance/Watt.

2.2.2.2 Memory Resources on Niagara

A block diagram of the Niagara memory subsystem is shown in Fig. 2.11 (left). On Niagara, the L1 instruction cache is 16 Kbyte, 4-way set-associative with a block size of 32 bytes. Niagara implements a random replacement scheme for area savings without incurring significant performance cost. The instruction cache fetches two instructions each cycle. If the second instruction is useful, the instruction cache has a free slot, which it can use to handle a line fill without stalling the pipeline. The L1 data cache is 8 Kbytes, 4-way set-associative with a line size of 16 bytes, and implements a write-through policy. To allow threads to continue execution past stores while maintaining the total store ordering (TSO) memory model, each thread has a dedicated, 8-entry store buffer. Even though the Niagara L1 caches are small, they significantly reduce the average memory access time per thread with miss rates in the range of 10%. Because commercial server applications tend to have large working sets, the L1 caches must be much larger to achieve significantly lower miss rates, so the Niagara designers observed that the incremental performance gained by larger caches did not merit the area increase. In Niagara, the four threads in a processor core are very effective at hiding the latencies from L1 and L2 misses. Therefore, the smaller Niagara level-one cache sizes are a good tradeoff between miss rates, area, and the ability of other threads in the processor core to hide latency.

Niagara was designed from the ground up to be a 32-thread CMP, and as such, employs a single, shared 3 MB L2 cache. This cache is banked 4 ways and pipelined to provide 76.8 GB/s of bandwidth to the 32 threads accessing it. In addition, the cache is 12-way associative to allow the working sets of many threads to fit into the cache without excessive conflict misses. The L2 cache also interleaves data across banks at a 64-byte granularity. Commercial server code has data sharing, which can lead to high coherence miss rates. In conventional SMP systems using discrete processors with coherent system interconnects, coherence misses go out over low-frequency off-chip buses or links, and can have high latencies. The Niagara design with its shared on-chip cache eliminates these misses and replaces them with low latency shared-cache communication. On the other hand, providing a single shared L2 cache

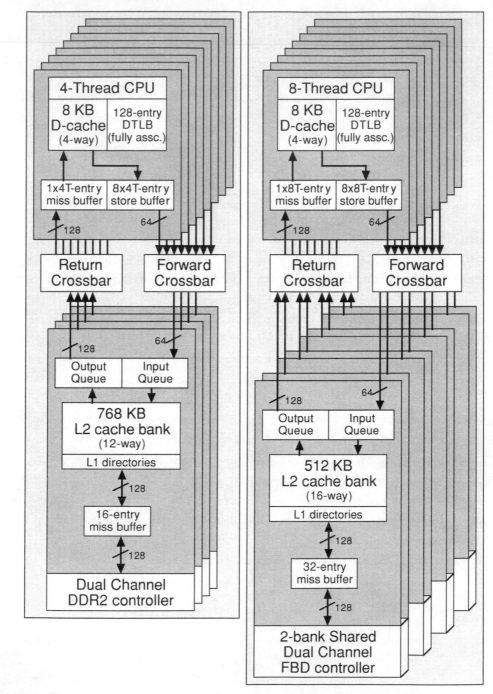

FIGURE 2.11: (Left) Niagara-1 memory hierarchy overview. (Right) Niagara-2 memory hierarchy overview.

implies that a slightly longer access time to the L2 cache will be seen by the processors, as the shared L2 cache cannot be located close to all of the processors in the chip. Niagara uses a crossbar to connect the processor cores and L2 cache, resulting in a uniform L2 cache access time. Unloaded latency to the L2 cache is 23 clocks for data and 22 clocks for instructions.

High off-chip bandwidth is also required to satisfy the L2 cache misses created by the multibank L2 cache. Niagara employs four separate memory controllers (one per L2 cache bank) that directly connect to DDR2 SDRAM memory DIMMs running at up to 200 MHz. Direct connection to the memory DIMMs allows Niagara to keep the memory latency down to 90 ns unloaded at 200 MHz. The datapath to the DIMMs is 128 bits wide (plus 16 bits of ECC), which translates to a raw memory bandwidth of 25.6 GB/s. Requests can be reordered in the Niagara memory controllers, which allow the controllers to favor reads over writes, optimize the accesses to the DRAM banks, and to minimize the dead cycles on the bi-directional data bus.

Niagara's crossbar interconnect provides the communication link between processor cores, L2 cache banks, and other shared resources on the CPU; it provides more than 200 Gbytes/s of bandwidth. A 2-entry queue is available for each source–destination pair, allowing the crossbar to queue up to 96 transactions in each direction. The crossbar also provides a port for communication with the I/O subsystem. Arbitration for destination ports uses a simple age-based priority scheme that ensures fair scheduling across all requestors. The crossbar is also the point of memory ordering for the machine.

Niagara uses a simple cache coherence protocol. The L1 caches are write through, with allocate on load and no-allocate on stores. L1 lines are either in valid or invalid states. The L2 cache maintains a directory that shadows the L1 tags. A load that missed in an L1 cache (load miss) is delivered to the source bank of the L2 cache along with its replacement way from the L1 cache. There, the load miss address is entered in the corresponding L1 tag location of the directory, the L2 cache is accessed to get the missing line and data is then returned to the L1 cache. The directory thus maintains a sharers' list at L1-line granularity. A subsequent store from a different or same L1 cache will look up the directory and queue up invalidates to the L1 caches that have the line. Stores do not update the local caches until they have updated the L2 cache. During this time, the store can pass data to the same thread but not to other threads; therefore, a store attains global visibility in the L2 cache. The crossbar establishes TSO memory order between transactions from the same and different L2 banks, and guarantees delivery of transactions to L1 caches in the same order.

2.2.2.3 Comparing Niagara with a CMP Using Conventional Cores

With the availability of real Niagara systems, a variety of real results have become publicly available. These benchmark results highlight both the performance and performance/Watt advantages of simple core CMPs. Figure 2.12(a) shows a comparison of SPECjbb 2005 results

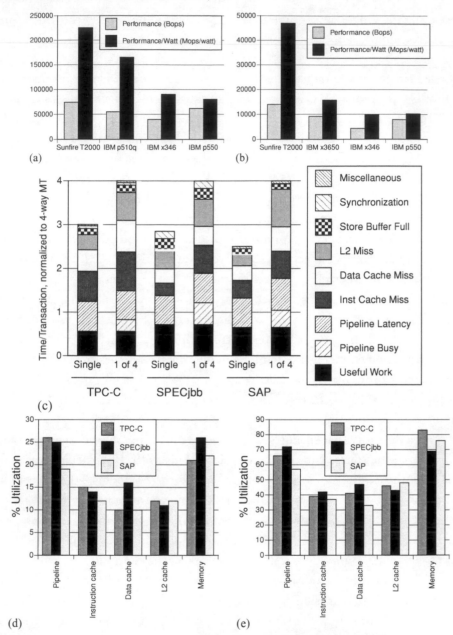

FIGURE 2.12: (a) SPECjbb 2005 performance; (b) SPECweb 2005 performance. (c) TPC-C, SPECjbb 2005, and SAP SD time breakdown within a thread, relative to the average execution time per task on a 4-thread system. The overall single-threaded times show slowdown from the multithreaded case, while each thread in a 4-thread system simply takes four times the average time per task, by definition. (d) 8-thread (1/processor) execution unit utilization; (e) 32-thread (4/processor) execution unit utilization.

between the Niagara-based SunFire T2000 and three IBM systems based on CMPs using more conventional superscalar POWER or ×86 cores: the IBM p510Q, IBM ×346, and IBM p550. The SunFire T2000 has nearly twice the performance/Watt of the closest system, the IBM p510Q, built from two 1.5 GHz Power5+ processors. The SunFire T2000 handily outperforms all three systems as well, despite the fact that the other systems all have two processor chips per box.

Similar performance and performance/Watt advantages of the Niagara can be seen for SPECweb 2005. Figure 2.12(b) shows the comparison between the SunFire T2000, IBM ×3650, IBM ×346, and IBM p550 on SPECweb 2005. For SPECweb 2005, the performance/Watt gap is even larger, with the SunFire T2000 having nearly three times the performance/Watt of the nearest competitor, the IBM ×3650, built from two dual-core 3.0 GHz Xeon 5160 processors. Again, the single processor chip SunFire T2000 outperforms the competing dual processor chip systems.

2.2.2.4 Niagara Performance Analysis and Limits

The large number of threads combined with limited speculation and a shallow pipeline allows Niagara to achieve excellent performance and performance/Watt on throughput workloads. Figure 2.12(c) shows that processing an entire database workload with only one thread running per-core takes three times as long as processing the workload with four threads running per core, even though the individual threads execute more slowly in that case. Note that adding multithreading does *not* achieve the optimal four times speedup. Instead, a comparison of the single-threaded and one-of-four thread time breakdowns allows us to see that destructive interference between the threads in the caches overcomes any interthread constructive cache interference and leads to increases in the portion of time that each of the multiple threads spends stalled on instruction cache, data cache, and L2 cache misses. There is also some time lost to contention for the shared pipeline, leading to an overall thread-vs.-thread slowdown of 33%. However, the multithreading allows much of the increased cache latency and nearly *all* of the pipeline latency to be overlapped with the execution of other threads, with the result being a threefold speedup from multithreading.

Similar results can be seen for SPECjbb 2005 and SAP SD. On both SPECjbb 2005 and SAP SD, the single thread is running at a higher efficiency than the database workload (CPIs of 4.0 and 3.8 are achieved for SPECjbb 2005 and SAP SD, respectively, compared to a CPI of 5.3 for the database workload), and as a result, the slowdown for each thread resulting from multithreading interference is larger, 40% for SPECjbb 2005 and 60% for SAP SD. As a result of the increased interference, the gains from multithreading visible in Fig. 2.12(c) are slightly lower for the two benchmarks, with SPECjbb 2005 showing a 2.85 times speedup, and SAP SD a 2.5 times performance boost.

A closer look at the utilization of the caches and pipelines of Niagara for the database workload, SPECjbb 2005, and SAP SD shows where possible bottlenecks lie. As can be seen from Figs. 2.12(d) and 2.12(e), the bottlenecks for Niagara appear to be in the pipeline and memory. The instruction and data caches have sufficient bandwidth left for more threads. The L2 cache utilization can support a modestly higher load as well. While it is possible to use multiple-issue cores to generate more memory references per cycle to the primary caches, a technique measured in [13], a more effective method to balance the pipeline and cache utilization may be to have multiple single-issue pipelines share primary caches. Addressing the memory bottleneck is more difficult. Niagara devotes a very large number of pins to the four DDR2 SDRAM memory channels, so without a change in memory technology, attacking the memory bottleneck would be difficult. A move from DDR2 SDRAM to Fully buffered DIMMs (FB-DIMM), a memory technology change already underway, is the key that will enable future Niagara chips to continue to increase the number of on-chip threads without running out of pins. Niagara 2, the second- generation version of Niagara, uses both multiple single-issue pipelines per core to attack the execution bandwidth bottleneck and FB-DIMM memories to address the memory bottleneck. The next section looks at the CMP design produced following these changes.

2.2.3   Example 3: The Niagara 2 Server CMP

Niagara 2 is the second-generation Niagara chip from Sun Microsystems. A block diagram of Niagara 2 is shown in Fig. 2.13. As with Niagara, Niagara 2 includes eight processor cores; however, each core doubles the thread count to eight, for a total of 64 threads per Niagara 2 chip. In addition, each core on Niagara 2 has its own floating-point unit, allowing Niagara 2 to address workloads with significant floating-point activity. The eight threads in Niagara 2 are partitioned into two groups of four threads. Each thread group has its own dedicated execution pipeline, although the two thread groups share access to a single set of data and instruction caches and the core's floating-point unit. The instruction and data caches are of the same size as on Niagara, 16 KB and 8 KB respectively, although the associativity on the instruction cache was increased to eight ways to accommodate the additional threads. As in Niagara, a dedicated 8-entry store buffer is provided per thread, doubling the total number of per-core store buffer entries in Niagara 2. The instruction TLB remains 64 entries on Niagara 2, but the data TLB is doubled in size to 128 entries. A hardware table walker was added in Niagara 2, which performs a hardware refill of the TLB on most TLB misses. The integer pipeline depth was increased slightly over Niagara, to eight stages, while the floating-point pipeline length was reduced to 12 stages. A 4-entry instruction buffer is provided for each thread and branches are predicted not taken in Niagara 2. A mispredicted branch causes the instruction buffer to be flushed and then a new fetch to refill the buffer from the branch target.

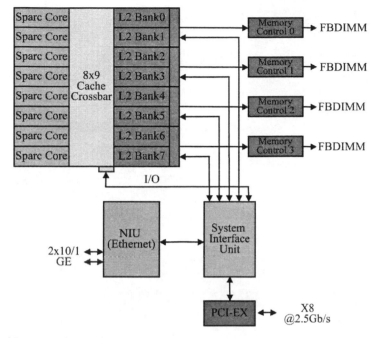

FIGURE 2.13: Niagara-2 block diagram.

The shared L2 cache in Niagara 2 was increased to 4 MB and 16-way set associativity. In addition, the L2 bank count was doubled to eight to support the increased load from the Niagara 2 processor cores. Niagara 2 retains four memory controllers, but each memory controller now connects to a dual-width FB-DIMM memory port, thereby greatly increasing the memory bandwidth while reducing the memory pin count. High-speed SERDES links, such as those employed in FB-DIMM interfaces, are the key that will enable future Niagara chips to continue to increase the number of on-chip threads without running out of memory pins. A block diagram of the Niagara 2 memory subsystem is shown in Fig. 2.11 (right). The I/O interface on Niagara 2 was changed from JBUS to an 8× PCI Express 1.0 port. Niagara 2 also has a greater level of integration over Niagara. Niagara 2 includes a pair of on-chip 10/1 Gb Ethernet ports with on-chip classification and filtering. The per-core cryptography units support a wider range of modular arithmetic operations as well as supporting bulk encryption. Ciphers and hashes supported include RC4, DES, 3DES, AES-128, AES-192, AES-256, MD5, SHA-1, and SHA-256. The cryptography unit in Niagara 2 is designed to process data at wire rates across the dual Ethernet ports. Niagara 2 is fabricated in TI's 65 nm process and is 342 mm^2. The use of FB-DIMM memory links and the high-level of SOC integration allows Niagara 2 to keep the processor cores fed using only 711 signal pins (1831 total pins). A die of Niagara 2 is shown in Fig. 2.14. Niagara 2 retains the same basic layout as Niagara, with four

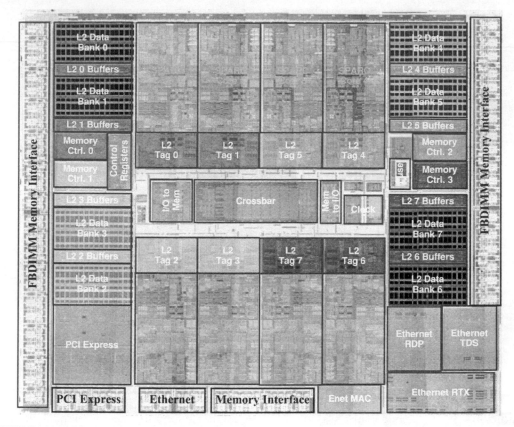

FIGURE 2.14: Niagara-2 die microphotograph.

cores on the top of the die, four on the bottom with the L2 tags and crossbar located between the cores. Four L2 data banks and their corresponding pair of memory controllers lie on the left side of the cores, with another four L2 data banks and corresponding memory controller pair on the right side of the cores. FB-DIMM I/O pads line the left and right edges of the die. The PCI-Express port and pads are located in the lower left of the die, while the Ethernet controller and MAC lie in the right corner of the die, connecting to the Ethernet pads located next to the PCI-Express pads. Filling in the spaces between the major blocks are clocks, test circuits (Fuse), memory-mapped control registers, and the interconnect from the I/O blocks to memory.

The goal of Niagara 2 was to achieve greater than twice the performance of Niagara at roughly the same power. To keep the power down, clocks are extensively gated throughout the Niagara 2 design. In addition, Niagara 2 has the ability to throttle issue from any of the threads and keeps speculation to a minimum. With greatly improved floating-point and SIMD instruction performance (Niagara 2 supports Sun's VIS 2.0 instructions directly in hardware) and much more significant system-on-a-chip (SOC) functions, including being able to run encrypted at wire rate, Niagara 2 is expected to address the needs of a even broader market space than Niagara.

2.2.4   Simple Core Limitations

Of course, there is no such thing as a free lunch, and simple core CMPs such as Niagara, Niagara 2, and Piranha do have some limitations.

As has already been discussed, low single-thread performance could be a potential disadvantage of the simple-core approach. Most large scale commercial applications, such as e-commerce, online transaction processing (OLTP), decision support systems (DSS), and enterprise resource planning (ERP) are heavily threaded, and even for nonthreaded applications there is a trend toward aggregating those applications on a common server pool in a grid computing fashion. Simple core CMPs like Piranha, and in particular multithreaded, simple core CMPs like Niagara, can have a significant advantage in both throughput performance and performance/Watt on these heavily threaded workloads. For workloads where the parallelism is low, however, the highest performance will be achieved by a CMP built from more complex cores capable of exploiting instruction-level parallelism from the small number of available software threads.

An additional possible limitation results from the area efficiency of the simple core. Because it is possible to place many more simple cores on the same die, the total thread count of the CMP can become quite large. This large thread count is an advantage when the workload has sufficient parallelism, but when coupled with the lower single-thread performance can become a liability when insufficient parallelism exists in the workload. A single Niagara chip has roughly the same number of threads as today's medium-scale symmetric multiprocessors. Going forward, future Niagara processors will likely include more threads per chip and support multiple chips in a single shared-memory system, leading to a very large number of active threads switched by hardware even in small, cost-effective systems. While many commercial applications have sufficient parallelism to be able to scale to several hundreds of threads, applications with more limited scalability will only be able to use a fraction of the threads in a future Niagara system. Of course, workloads consisting of many of these more limited scalability applications can be multiprogrammed on Niagara systems under control of the operating system. In addition, Niagara, along with many of the more recent processors, has hardware support for virtualization, and multiple operating systems (each referred to as a "guest" OS) can be run on a single Niagara system, with each guest operating system running their own set of application workloads. As single systems become capable of running what used to require multiple dedicated mainframes, this ability to consolidate multiple workloads, each running under their own guest operating system, fully protected from the effects of other guest operating systems, will become extremely important.

2.3   GENERAL SERVER CMP ANALYSIS

The case studies of Piranha and Niagara provide interesting examples of CMPs that were well-designed for use with server workloads, but they also raise further questions, since CMPs offer such a large design space. Examples include: Given a target chip area, how should the area be allocated between cores and caches? How many threads per core? How complex of a pipeline should each core have? How should cache be divided between L1 and L2? How should everything be connected? There are so many possible variations that it is an interesting question to try and determine how these different sorts of design choices interact to affect performance. One way to do this is simply to evaluate a wide variety of design points and look at the trends that emerge.

2.3.1   Simulating a Large Design Space

Using the Niagara design as a baseline, it is possible to investigate the CMP design space in detail. By exploring several of Sun Microsystem's UltraSPARC chip design databases, the authors of [13] determined the area impact of the architectural components when they are modified to enable fine-grain multithreading. From this, they derived a thread-scalable fine-grained multithreaded processor core area model which correlates well with actual and projected UltraSPARC processor areas from 130 nm to 45 nm silicon process generations. Using a CMP built with a single shared level-two cache as a baseline, they were able to simulate four server applications across a wide variety of CMPs built on a "fixed" 400 mm^2 die (a size chosen because historically most server vendors have chosen to manufacture high-end chips at or near this size) in several process technologies.

Figure 2.15 illustrates and Table 2.2 describes the variety of high-level CMT configurations; all the gray components were varied in the study. The processor cores can utilize either in-order scalar or superscalar integer datapaths (IDPs). The number of threads per IDP and the number of independent IDPs within each core (where a "core" is defined as a pair of L1 caches and all associated processor pipelines) were both varied. In the scalar processor design, threads were statically assigned to an IDP, as this avoids the superlinear area impact of being able to issue instructions from any of the threads on a core to any of the IDPs. All cache sizes and set associativities (SA) could vary. Instruction caches and data caches were always identical in size or differed by a factor of 2X, but no more. The primary caches ranged from 8 KB to 128 KB with SA ranging from direct mapped to 8-way. Small instruction buffers for each thread decoupled the front end of each IDP from the shared primary instruction cache. The memory and cache subsystems were fully modeled with queuing, delaying, and occupancy. The actual

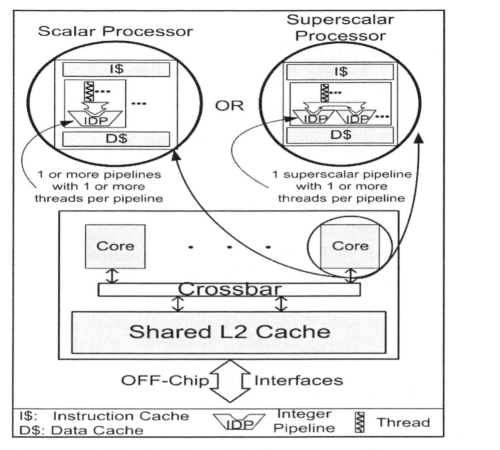

FIGURE 2.15: A high-level functional diagram of the CMT design space. The gray components are varied and described in Table 2.2.

RAS/CAS cycles for the DRAM accesses were modeled along with all the various buffers and queues. The number of processor cores and sizes of the caches were determined by the area model for a given silicon process technology, keeping die size constant across all possible configurations. More details on the area scaling model are available in [13].

To drive this large selection of CMPs, the authors used SPEC JBB 2000, TPC-C, TPC-W, and XML Test server benchmarks to assess the CMT's performance. SPEC JBB emulates a three-tier system emphasizing the Java server-side performance of middleware business logic [27]. TPC-C is an online transaction processing benchmark based on an order-entry system [25], and only the server side was used in this study. TPC-W is a transactional web benchmark that simulates the activities of a business- oriented transactional web server [28]. XML Test is a multithreaded XML processing test developed at Sun Microsystems [29]. XML Test performs both streaming and tree-building parsing, which replicate application servers that provide web

TABLE 2.2: CMT design space parameters

FEATURE	DESCRIPTION
CPU	In-order scalar or superscalar
Issue Width	Scalar, 2-way and 4-way superscalar
Pipeline Depth	8 stages
Integer Datapath Pipelines	1–4 IDPs or Integer ALUs
L1 D & I Cache	8 KB–128 KB, 16 (D) & 32 (I) Byte lines
L1 D & I Cache Set Assoc.	Direct-mapped, 2-, 4-, or 8-way
L1 D & I Cache Policies	Write-through, LRU-based replacement
Clock Frequency	1/3–1/2 of the maximum ITRS clock frequency [23]
Multithreading	1–32 threads/core
L2 Cache	1 MB–8 MB, 128 byte lines, (8 or 16), coherent, banked inclusive, shared, unified, critical word first, 25 cycle hit time (unloaded)
Main Memory	Fully buffered DIMMs with 4/8/16 dual channels, 135 cycle latency (unloaded)

services and simultaneously process XML documents. Unlike SPEC JBB, XML Test is a single-tier system benchmark; the test driver is integrated into the worker thread.

These benchmarks do not exhibit multiphase execution, so recording contiguous streams of instruction on a per thread basis can capture the complete system performance, the overall benchmark characteristics, and the instruction mix. In contrast, benchmarks like SPEC CPU2000 require sampling techniques to capture the various phases of execution [30]. The study looked at commercial-grade configurations of the various benchmarks. SPEC JBB used the J2SE 1.4 JVM with a 2 GB heap running on Solaris 9 with 16 warehouses to collect a 16-processor instruction trace file. XML Test used the J2SE 1.5 JVM, but with a 2.5 GB heap for a 16-processor trace file. TPC-C required 3000 warehouses with a 28 GB SGA and 176 9 GB disks coupled with commercial database management and volume manager software running on Solaris 9. For both TPC-C and TPC-W, the clients and servers were simulated, but only the server instruction traces were used in their study. TPC-W was configured to support up to 10,000 users. The database was built on 28 9 GB disks coupled with commercial database management and volume manager software running on Solaris 9. The application server used JDK 1.4.x, while JDK 1.3.x was used for the image server, payment gateway emulator, and the SSL components. Fixed processor sets were used to isolate the application servers from the rest of the simulation, allowing the instruction streams from only the application server processor set to be gathered.

Each trace, captured on a real machine running the benchmark, contained several billion instructions per process thread in steady state. The traces were collected during the valid measurement time after the benchmarks had ramped up and completed the benchmark specified warm-up cycle, as on real hardware. They observed significant variation in benchmark performance during the ramp-up period, but little variation once in steady state, which was also observed in [31]. All benchmarks were highly tuned, with less than 1% system idle time, and showed negligible performance variability during the measurement period. After benchmark capture, the traces were used to drive a relatively fast system simulator that was parameterized to be able to reconfigure into the wide variety of configurations needed for the study.

2.3.2 Choosing Design Datapoints

Table 2.3 summarizes the parameter ranges that were investigated for 90 nm CMPs. The maximum primary cache capacities are shown as a single value or as X/Y if the maximum is asymmetric, where one L1 cache is larger than the other. For this latter case, the set associativity of the larger cache in the asymmetric pair remained low to further constrain the area. Finally, four secondary cache sizes for each of the 21 core configurations, corresponding to approximately 25%, 40%, 60%, and 75% of the CMT area, were simulated. To prevent the DRAM bandwidth from becoming a bottleneck, their study used an aggressive but achievable design with eight dual FB-DIMM memory channels connecting to eight L2 cache banks.

The in-order scalar and superscalar cores utilize fully pipelined integer and floating-point datapaths, with each datapath capable of executing one instruction per cycle. Each processor core consists of one to four integer datapath pipelines (IDPs or integer ALUs). Up to eight hardware threads are supported per IDP within the processor core, while up to eight hardware threads are supported per superscalar processor core. The nomenclature used to label the scalar cores is NpMt, where N is the number of IDPs in the core, and M is the total number of hardware threads supported by the core. The scalar cores were differentiated from the superscalar cores by labeling them NsMt, where N denotes the issue width of the superscalar processor. Each scalar integer pipeline can only execute instructions from a statically assigned pool of M/N threads, whereas the superscalar pipelines can issue instructions from any of M threads. Each core contains a single-ported primary data and instruction cache shared between the IDPs, sized from 8 KB up to the values shown in Table 2.3.

TABLE 2.3: CMT design space parameters segmented (alternating gray areas) to indicate major core configuration groups. All L2 cache configurations are used with all core configurations per class

CORE CONFIG	NUMBER OF IDPS	NUMBER OF THREADS	MAX L1 SIZE (KB)	L2 CACHE (MB, SA)	NUMBER OF PROCESSORS	AGGREGATE THREADS
1p2t	1	2	32		5–20	10–40
1p4t	1	4	32		5–17	20–68
1p8t	1	8	64		3–14	24–112
2p2t	2	2	32/64		4–16	8–32
2p4t	2	4	64		3–14	12–112
2p8t	2	8	64/128	1.5, 12	3–12	24–96
2p16t	2	16	128		2–9	32–144
3p3t	3	3	64		3–13	9–39
3p6t	3	6	64/128	2.5, 10	3–11	18–66
3p12t	3	12	128		2–9	24–108
3p24t	3	24	128		1–6	24–144
4p8t	4	8	64/128	3.5, 14	2–9	16–72
4p16t	4	16	128		2–7	32–112
2s1t	2	1	64		4–11	4–11
2s2t	2	2	64	4.5, 18	4–10	8–20
2s4t	2	4	64		3–9	12–36
2s8t	2	8	64		2–7	16–56
4s1t	4	1	64		2–7	2–7
4s2t	4	2	64		2–6	4–12
4s4t	4	4	64		2–5	8–20
4s8t	4	8	64		1–4	8–32

2.3.3 Results

The commercial server applications exhibited a range of low to moderate ILP and high cache miss rates similar to the observations in [9]. Using a single thread per pipeline provides no hardware mechanism for latency tolerance and results in low processor utilization, or "under-threading." On the other hand, too many active threads can lead to an "overthreaded" core with a fully utilized integer datapath pipeline (IDP) and performance that is insensitive to primary cache capacity or set associativity. The goal of their study was to find the right design balance that optimized aggregate IPC of the entire CMP.

Historically, the goal of optimizing the processor core was to squeeze out every last percent of performance that could be achieved with reasonable area costs. In the CMP design space, this is a local optimization that is not likely to yield high aggregate performance. This is exemplified by the aggregate IPC results for the 2p4t core configuration shown in Fig. 2.16. The top two lines are the aggregate IPCs (AIPCs) for a particular cache configuration and the bottom two lines are the corresponding average core IPCs. C1 represents the 2p4t configuration with the best core IPC (64 KB data and instruction cache), but the corresponding full-chip CMP built from C1 has an AIPC that underperforms due to the small number of cores that can be fit on the die. On the other hand, C2 is a "mediocre" 2p4t configuration with only a 32 KB data and instruction cache, but it has the best AIPC by maximizing the number of cores on the CMP for any given secondary cache size, as indicated in Fig. 2.16. Figure 2.16 also illustrates that too many cores on the CMP can degrade overall performance. As the area of the CMP devoted to level-two cache decreases from 2.5 MB down to 1.5 MB, the AIPC of

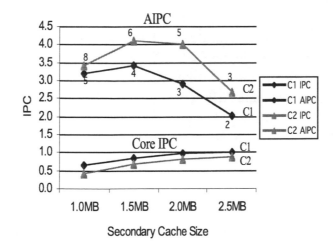

FIGURE 2.16: CMT TPC-C core and aggregate IPC for the 2p4t CMT configuration (a smaller size than measured). C1 has the best average core IPC. C2 has the best aggregate IPC by using more cores on the die. The number of cores for each CMT is labeled next to the upper pair of lines.

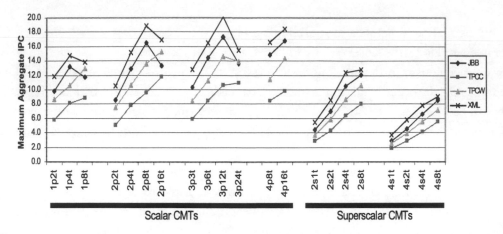

FIGURE 2.17: Medium-scale aggregate IPC for each CMT configuration and all benchmarks.

the CMP increases from the higher thread count that results from being able to fit more cores on the die. However, decreasing the level-two cache size from 1.5 MB to 1.0 MB results in a drop in performance, resulting from the combined working sets of the threads thrashing in the now too small level-two cache. As both the total number of cores that can be fit on the chip and the performance of each of those cores are strongly dependent on the amount of on-chip secondary cache, it is important to balance processing and cache needs.

The best results for each core configuration and all of the benchmarks used in this study are presented in Fig. 2.17 for the 90 nm CMPs. This figure provides the maximum AIPC (y-axis) across all cache configurations for all pipeline/thread configurations (x-axis). The number of cores and cache configurations that yield the AIPC in Fig. 2.17 is provided in Table 2.4 for each pipeline/thread configuration. The CMPs are clustered by pipeline and pipeline architecture, scalar vs. superscalar.

Table 2.4 shows the maximum AIPC for SPEC JBB, TPC-C, TPC-W, and XML Test for the 90 nm CMPs. This table lists the best configuration for each core configuration and highlights the overall best CMP configuration in black boxes.

2.3.4   Discussion

The results of this study showed that augmenting CMPs with fine-grain multithreading is crucial to increasing the performance of commercial server applications. While multiple processor cores can exploit TLP, fine-grain multithreading is also necessary to alleviate the otherwise poor core utilization for these applications. However, fine-grain multithreading runs into two limits. First, the addition of too many threads to a core resulted in a saturated integer pipeline

TABLE 2.4: Maximum AIPC for medium-scale CMTs for SPEC JBB, TPC-C, TPC-W, and XML test. L1 data/instruction sizes are in KB. L2 statistics are given as size in MB/set associativity.

CORE CONFIG	SPEC JBB 2000				TPC-C				TPC-W				XML TEST			
	L1	L2	CORES	AIPC	L1	L2	CORES	AIPC	L1	L2	CORES	AIPC	L1	L2	CORES	AIPC
1p2t	16/32	1.5/12	20	9.8	16/32	2.5/10	16	5.8	16/32	1.5/12	20	8.6	16/32	1.5/12	17	14.8
1p4t	16/32	1.5/12	17	13.2	16/32	2.5/10	14	8.2	16/32	1.5/12	17	10.6	16/32	1.5/12	17	14.8
1p8t	16/32	2.5/10	12	11.7	32/32	1.5/12	14	8.9	32/32	1.5/12	14	13.0	16/32	1.5/12	14	13.8
2p2t	16/32	1.5/12	16	8.6	16/32	1.5/12	16	5.1	16/32	1.5/12	16	7.5	16/32	1.5/12	16	10.5
2p4t	32/32	1.5/12	14	12.9	32/32	2.5/10	12	7.8	32/32	1.5/12	14	10.6	16/32	1.5/12	14	15.2
2p8t	16/32	1.5/12	12	16.5	32/32	2.5/10	9	9.5	32/32	1.5/12	12	13.6	32/32	1.5/12	12	18.9
2p16t	32/64	2.5/10	7	13.3	64/64	2.5/10	7	**11.8**	64/64	1.5/12	9	**15.2**	32/64	1.5/12	9	16.9
3p3t	32/32	1.5/12	13	10.3	32/32	2.5/10	10	5.9	32/32	1.5/12	13	8.5	16/32	1.5/12	13	12.5
3p6t	32/32	1.5/12	11	14.4	32/32	2.5/10	9	8.5	32/32	1.5/12	11	11.3	32/32	1.5/12	11	16.5
3p612t	32/64	1.5/12	9	**17.3**	32/64	2.5/10	7	10.7	64/64	1.5/12	9	14.6	32/64	1.5/12	9	**20.1**
3p24t	32/64	2.5/10	5	13.6	32/64	2.5/10	5	10.9	32/64	1.5/12	6	14.0	32/64	1.5/12	6	15.5
4p8t	32/32	1.5/12	9	14.9	32/32	2.5/10	7	8.5	64/64	1.5/12	9	11.5	16/32	1.5/12	9	16.6
4p16t	32/64	1.5/12	7	16.8	32/64	2.5/10	5	9.8	64/64	1.5/12	7	14.4	32/64	1.5/12	7	18.5
2s1t	64/64	1.5/12	11	4.4	64/64	1.5/12	11	2.8	64/64	1.5/12	11	3.7	64/64	1.5/12	11	5.5
2s2t	64/64	1.5/12	10	7.0	64/64	1.5/12	10	4.3	64/64	1.5/12	10	5.8	64/64	1.5/12	10	8.5
2s4t	64/64	1.5/12	9	10.5	64/64	1.5/12	9	6.4	64/64	1.5/12	9	8.7	64/64	1.5/12	9	12.4
2s8t	64/64	1.5/12	7	12.1	64/64	1.5/12	7	8.1	64/64	1.5/12	7	10.6	64/64	1.5/12	7	12.7
4S1t	64/64	1.5/12	7	2.9	64/64	1.5/12	7	1.9	64/64	1.5/12	7	2.6	64/64	1.5/12	7	3.7
4S2t	64/64	1.5/12	6	4.5	64/64	1.5/12	6	2.9	64/64	1.5/12	6	3.9	64/64	1.5/12	6	5.8
4S4t	64/64	1.5/12	5	6.6	64/64	1.5/12	5	4.1	64/64	1.5/12	5	5.6	64/64	1.5/12	5	7.8
4S8t	64/64	1.5/12	4	8.5	64/64	1.5/12	4	5.5	64/64	1.5/12	4	7.2	64/64	1.5/12	4	9.1

that wasted silicon to support threads which added little to performance. In the study, this saturation occurred with about eight threads per integer pipeline for scalar cores. Second, a CMP built with too many total threads for the secondary cache size could end up saturating the memory bandwidth with secondary cache misses, as the aggregate working set can overflow the secondary cache. Memory saturation occurred in the study primarily with configurations that had the smallest secondary cache size (occupying 24–28% of the CMP area) and eight or more threads per core. For the server applications, aggregate IPC tended to be optimized by a processor-centric design, requiring only 25–40% of the area devoted to the shared secondary cache. For the primary caches, a larger primary instruction cache than the primary data cache was always the best policy. Surprisingly, high primary cache set associativity was not required for these applications, even with more threads than set associative ways.

For a given primary data and instruction cache configuration, the performance difference based on set associativity varied less than 3% for the best aggregate IPC configurations, as long as the caches were at least 2-way set associative. The best performing configurations required enough threads and primary cache to bring the pipeline utilization up to the 60–85% range, as the area costs for adding additional pipelines and threads per pipeline is much smaller than adding an additional core. The best configuration was with 3 pipelines and 12 threads per core for Spec JBB amd XML Test, while 2 pipelines and 16 threads per core performed best for TPC-C and TPC-W. In addition, the best performing CMP configuration was highly dependent on a step function of the number of cores that can be squeezed on the die, allowing a CMP composed of slightly lower performance cores to yield superior aggregate performance by employing more of those cores. As a corollary to this step function regarding core size, processor cores with smaller primary caches were favored, even without penalizing the larger caches with additional latency, as the smaller-cache cores maximized the number of on-chip cores.

Interestingly enough, the worst performing configurations for SPEC JBB and XML Test included both "underthreaded" and "overthreaded" configurations, while the worst performing configurations for TPC-C and TPC-W were always "underthreaded" configurations. This matches intuition, as the low IPC of a single TPC-C or TPC-W thread makes underthreading more detrimental, while the more moderate IPC of a single SPEC JBB or XML Test thread makes it more susceptible to both underthreading and overthreading.

Finally, as expected for these heavily threaded workloads, scalar CMP variants with four or more threads readily outperformed nearly all of the superscalar CMP configurations given the constant die size constraint. Comparing performance just within the superscalar designs, 2-way superscalar configurations outperformed all 4-way superscalar configurations with the same number of threads.

Not surprisingly, CMPs are an effective way to design large processor chips that perform well on heavily threaded, throughput-oriented workloads. The primary issue that must be considered in the design of CMPs for these applications is properly balancing numbers of cores, numbers of threads per core, the sizes of caches, and off-chip memory bandwidth. While finding the best balance can be difficult, it is a relatively straightforward task. We now move to the much more difficult task of speeding up latency-bound applications using the multiple cores within a CMP.

REFERENCES

[1] L. Barroso, J. Dean, and U. Hoezle, "Web search for a planet: the architecture of the Google cluster," *IEEE Micro.*, Vol. 23, No. 2, pp. 22–28, Mar.–Apr. 2003.

[2] K. Olukotun, B. A. Nayfeh, L. Hammond, K. Wilson, and K. Chang, "The case for a single chip multiprocessor," in *Proc. 7th Int. Conf. Architectural Support for Programming Languages and Operating Systems (ASPLOS-VII)*, Cambridge, MA, Oct. 1996, pp. 2–11.

[3] S. Kapil, "UltraSPARC Gemini: dual CPU processor," in *Hot Chips 15*, Stanford, CA, Aug. 2003. http://www.hotchips.org/archives/

[4] T. Maruyama, "SPARC64 VI: Fujitsu's Next Generation Processor," in *Microprocessor Forum*, San Jose, CA, Oct. 2003.

[5] C. McNairy and R. Bhatia, "Montecito: the next product in the Itanium Processor Family," in *Hot Chips 16*, Stanford, CA, Aug. 2004. http://www.hotchips.org/archives/

[6] C. Moore, "POWER4 system microarchitecture," in *Microprocessor Forum*, San Jose, CA, Oct. 2000.

[7] L. A. Barroso, K. Gharachorloo, R. McNamara, A. Nowatzyk, S. Qadeer, B. Sano, S. Smith, R. Stets, and B. Verghese, "Piranha: a scalable architecture based on single-chip multiprocessing," in *Proc. 27th Int. Symp. Computer Architecture (ISCA-27)*, Vancouver, BC, Canada, June 2000, pp. 282–293.

[8] J. Lo, L. Barroso, S. Eggers, K. Gharachorloo, et al. "An analysis of database workload performance on simultaneous multithreaded processors," in *Proc. 25th Annu. Int. Symp. Computer Architecture (ISCA-25)*, Barcelona, Spain, June 1998, pp. 39–50.

[9] S. Kunkel, R. Eickemeyer, M. Lip, and T. Mullins, "A performance methodology for commercial servers," *IBM J. Res. Dev.*, Vol. 44, No. 6, Nov. 2000, pp. 851–872.

[10] T. Agerwala and S. Chatterjee, "Computer architecture: challenges and opportunities for the next decade," *IEEE Micro*, Vol. 25, No. 3, pp. 58–69, May/June 2005.

[11] P. Kongetira, K. Aingaran, and K. Olukotun, "Niagara: a 32-way multithreaded SPARC processor," *IEEE Micro*, Vol. 25, No. 2, Mar./Apr. 2005, pp. 21–29.

[12] J. Clabes, J. Friedrich, and M. Sweet, "Design and implementation of the POWER5™ Microprocessor," in *ISSCC Dig. Tech. Papers*, San Francisco, CA, Feb. 2004, pp. 56–57.

[13] J. D. Davis, et al. "Maximizing CMT throughput with Mediocre cores," in *Proc. 14th Int. Conf. Parallel Architectures and Compilation Techniques*, St. Louis, MO, Sept. 2005, pp. 51–62.

[14] D. Marr, "Hyper-threading technology in the Netburst Microarchitecture," in *Hot Chips XIV*, Stanford, CA, Aug. 2002. http://www.hotchips.org/archives/

[15] A. Agarwal, J. Kubiatowicz, D. Kranz, B.-H. Lim, D. Yeung, G. D'Souza, and M. Parkin, "Sparcle: an evolutionary processor design for large-scale multiprocessors," *IEEE Micro*, Vol. 13, No. 3, June 1993, pp. 48–61.

[16] J. Laudon, A. Gupta, and M. Horowitz, "Interleaving: a multithreading technique targeting multiprocessors and workstations," in *Proc. 6th Int. Symp. Architectural Support for Parallel Languages and Operating Systems (ASPLOS-VI)*, San Jose, CA, Oct. 1994, pp. 308–318.

[17] D. Tullsen, S. Eggers, and H. Levy, "Simultaneous multithreading: maximizing on-chip parallelism," in *Proc. 22nd Annu. Int. Symp. Computer Architecture (ISCA-22)*, Santa Margherita Ligure, Italy, June 1995, pp. 392–403.

[18] S. Naffziger, T. Grutkowski, and B. Stackhouse, "The implementation of a 2-core multi-threaded Itanium® Family Processor," in *Proc. of IEEE International Solid-State Circuits Conference (ISSCC)*, San Francisco, CA, Feb. 2005, pp. 182–183.

[19] R. L. Sites and R. T. Witek, *Alpha AXP Architecture Reference Manual*, 2nd edition. Burlington, MA: Digital Press, 1995.

[20] N. P. Jouppi and S. Wilton, "Tradeoffs in two-level on-chip caching," in *21st Annu. Int. Symp. Computer Architecture (ISCA-21)*, Chicago, IL, Apr. 1994, pp. 34–45.

[21] Transaction Processing Performance Council. TPC Benchmark B, Standard Specification Revision 2.0, June 1994.

[22] Transaction Processing Performance Council. TPC Benchmark D (Decision Support) Standard Specification Revision 1.2, Nov. 1996.

[23] L. A. Barroso, K. Gharachorloo, and E. Bugnion. "Memory system characterization of commercial workloads," in *25th Annu. Int. Symp. Computer Architecture*, Barcelona, Spain, June 1998, pp. 3–14.

[24] M. Rosenblum, E. Bugnion, S. Herrod, and S. Devine, "Using the SimOS machine simulator to study complex computer systems," *ACM Trans. Model. Comput. Simul.*, Vol. 7, No. 1, pp. 78–103, Jan. 1997.

[25] Transaction Processing Performance Council. TPC Benchmark C, Standard Specification Revision 3.6, Oct. 1999.

[26] J. Laudon, "Performance/Watt: the new server focus," in *Proc. Workshop on Design, Architecture, and Simulation of Chip Multiprocessors*, Barcelona, Spain, Nov. 2005.

[27] Standard Performance Evaluation Corporation, SPEC, http://www.spec.org, Warrenton, VA.

[28] Transaction Processing Performance Council, TPC, http://www.tpc.org, San Francisco, CA.

[29] "XML Processing Performance in Java and .Net," http://java.sun.com/performance/reference/whitepapers/XML_Test-1_0.pdf

[30] T. Sherwood, S. Sair, and B. Calder, "Phase tracking and prediction," in *30th Annu. Int. Symp. Computer Architecture*, San Diego, CA, June 2003, pp. 336–347.

[31] A. R. Alameldeen and D. A. Wood, "Variability in architectural simulations of multi-threaded workloads," in *9th Int. Symp. High Performance Computer Architecture (HPCA)*, Anaheim, CA, Feb. 2003.

CHAPTER 3

Improving Latency Automatically

While high overall throughput of many essentially unrelated tasks is often important, there are still many important applications whose performance is measured in terms of the execution latency of individual tasks. Most desktop processor applications still fall in this category, as users are generally more concerned with their computers responding to their commands as quickly as possible than they are with its ability to handle many commands simultaneously, although this situation is changing slowly over time as more applications are written to include many "background" tasks, such as continuous spell checking. Users of many large, computation-bound applications, such as most simulations and compilations, are typically also more interested in how long the programs take to execute than in executing many in parallel.

Multiprocessors can be used to speed up these types of applications, but it normally requires effort on the part of programmers to break up each long-latency thread of execution into a large number of smaller threads that can be executed on many processors in parallel, since automatic parallelization technology has typically only functioned well on FORTRAN programs describing dense-matrix numerical computations and other, similar types of regular applications. Historically, communication between processors was generally slow in relation to the speed of individual processors, so it was critical for programmers to ensure that threads running on separate processors required only minimal communication between each other. Because communication reduction is often difficult, only a small minority of users bothered to invest the time and effort required to parallelize their programs in a way that could achieve speedup, and so these techniques were only taught in advanced, graduate-level computer science courses. In most cases programmers found that it was just easier to wait for the next generation of uniprocessors to appear and speed up their applications for "free" instead of investing the effort required to parallelize their programs. As a result, multiprocessors had a hard time competing against uniprocessors except for the most demanding workloads, where the target performance simply exceeded the power of the fastest uniprocessors available by orders of magnitude.

CMPs greatly simplify the problems traditionally associated with parallel programming to the point where it can largely be automated. While previously it was necessary to minimize communication between independent threads to an *extremely* low level, because each communication could require hundreds or even thousands of processor cycles, within a CMP with a shared on-chip cache memory each communication event typically takes just a handful of processor cycles. With latencies like this, it is now feasible to perform fairly extensive automatic program transformations in a compiler and/or at runtime to utilize two or more processors together on a single, latency-critical application. With conventional multiprocessors, this was simply not possible because uniprocessor programs are usually not designed with parallel execution in mind. If they are arbitrarily broken up and parallelized, one is virtually assured that a few critical communication paths will limit performance so much that speedups won't be possible. In contrast, the tightly coupled communication in a CMP means that speedups are now possible even when significant quantities of communication between threads are required. This can be leveraged to make manual parallelization of applications significantly easier (using techniques such as those described in Chapter 4), but just as importantly it now makes *automated* parallelization of nominally uniprocessor codes a feasible option for a *much* wider array of programs. To move beyond the limited set of dense-matrix applications that can already be parallelized automatically by compilers (and used in the analysis in Chapter 1), a CMP will typically require some additional hardware support.

3.1   PSEUDO-PARALLELIZATION: "HELPER" THREADS

The simplest way to use parallel threads within a CMP to improve the performance of a single thread is to have one or more "helper" threads perform work on behalf of the main thread in an effort to accelerate its performance. These "helper" threads perform work speculatively in order to compute key values ahead of time for the main thread or to start long-latency operations early. Two problems that have been addressed using this technique are making hard-to-predict branch predictions early [1] and prefetching irregular data into on-chip caches [2–4]. In most cases, hardware branch prediction methods and prefetch units perform fairly well, but for some very irregular codes it can be advantageous to actually compute branch directions or start prefetches before the main thread actually needs them. The remainder of this section focuses on the more extensively studied technique of using helper threads for prefetching.

To prefetch data for the main thread, the helper thread(s) run ahead of the main thread using copies of the thread that have been reduced to omit all code not essential to their assigned task, allowing them to speed ahead of the main thread. Since these threads just prefetch data into the nearest cache shared by all threads, the definition of "essential" is fairly loose. In order to avoid polluting the cache with useless data, the prefetches generated by the thread should generally have the same addresses as the ones in the main thread. However, some corners may be cut if they only occasionally result in spurious prefetches or missing prefetch opportunities. These errors will only affect performance somewhat, and will not actually result in incorrect execution. If the threads are selected well and properly synchronized with the main thread, they can significantly reduce the number of cache misses incurred by the main thread beyond the shared cache. Since modern processors can wait for hundreds of cycles on a miss to main memory, the potential savings from eliminating these misses can be quite significant for some applications.

In practice, the amount of speedup that can be attained using prefetching helper threads is usually fairly modest. The first problem is that only certain types of single-threaded programs can be sped up with these techniques. Most integer applications have fairly modest memory footprints, and therefore have few cache misses to eliminate. Many of those with large memory footprints, such as databases, are fairly easy to parallelize into true threads, which are always a better choice than "helper" threads if it is possible to create them. On the floating-point side, many applications are easily parallelizable or have fairly regular access patterns that can be prefetched using hardware mechanisms or occasional software prefetch instructions right in the main thread. As a result of these fundamental application characteristics, the selection of applications that can really be helped by these techniques is fairly limited. The second problem is that very tight synchronization and/or thread fork/join is needed between the main thread and its helpers in order to keep them the proper distance ahead of the main thread. Too far ahead, and they will cause cache thrashing by prefetching data and then replacing it with subsequent prefetches before the main thread can even use it. On the other hand, if they are not far enough ahead they might not be able to prefetch cache lines in time. Inserting just enough synchronization to keep these threads properly paced without slowing down the main thread significantly is still an active area for research.

Among the published literature, only a handful of applications, such as the linked-list intensive *mcf* from SPEC CPU2000 [5], have shown benefit from helper threads. [2] and [3], which both implemented helper threads on real hardware, saw less than 5% speedup from all but a few select applications ([3] managed to get a 22% boost with *mcf*, for example). In many cases, performance even slowed by a few percent, as the synchronization overhead between the threads overcame any potential benefit from the prefetching. Only studies that have assumed additional hardware to allow very fast interthread synchronizations, such as the hardware fork-join instructions in [4], have been able to demonstrate more impressive speedups (up to about 50% on the best application in that study, for example, although generally lower). Mechanisms such as this are possible on a CMP, but they may not be worthwhile if they only accelerate the performance of a very limited selection of applications.

3.2   AUTOMATED PARALLELIZATION USING THREAD-LEVEL SPECULATION (TLS)

Ultimately, in order for the parallel threads within a CMP to allow truly significant performance gains when accelerating single-threaded applications, the programs need to be parallelized so that all cores may *independently* complete computations from different portions of the program. Of course, splitting up a single-threaded program into multiple independent threads while still computing the correct result is considerably more challenging than simply creating "helper" threads to perform prefetches (which can affect performance, but never correctness). The remainder of this chapter discusses one CMP-based technique that is particularly useful for automating parallelization: *thread-level speculation* (TLS).

TLS is a technique that can automatically parallelize an existing uniprocessor program across the cores of a CMP, eliminating the need for programmers to explicitly divide an original program into independent threads. TLS takes the sequence of instructions run during an existing uniprocessor program and arbitrarily breaks it into a sequenced group of threads that may be run in parallel on a multiprocessor. To ensure that each program executes the same way that it did originally, hardware must track all interthread dependences. When a "later" thread in the sequence causes a true dependence violation by reading data too early, the hardware must ensure that the misspeculated thread—or at least the portion of it following the bad read—re-executes with the proper data. This is a considerably different mechanism from the one used to enforce dependences on conventional multiprocessors. There, synchronization is inserted so that threads reading data from a different thread will stall until the correct value has been written. This process is complex because it is necessary to determine *all* possible true dependences in a program before synchronization points may be inserted.

Speculation allows parallelization of a program into threads even without prior knowledge of where true dependences between threads may occur. All threads simply run in parallel until a true dependence is detected while the program is executing. This greatly simplifies the parallelization of programs because it eliminates the need for human programmers or compilers to statically place synchronization points into programs by hand or at compilation time. All places where synchronization would have been required are simply found dynamically when true dependences actually occur. As a result of this advantage, uniprocessor programs may be *obliviously* parallelized in a speculative system. While conventional parallel programmers must constantly worry about maintaining program correctness, compilers (or programmers) parallelizing code for a speculative system can focus solely on achieving maximum performance. The speculative hardware will ensure that the parallel code always performs the same computation as the original sequential program.

Since parallelization by speculation dynamically finds parallelism among program threads at runtime, it does not need to be as conservative as conventional parallel code. In many programs, there are many *potential* dependences that may result in a true dependence, but where dependences seldom if ever actually occur during the execution of the program. A speculative system may attempt to run the threads in parallel anyway, and only back up the later thread if a dependence actually occurs.

On the other hand, a system dependent on synchronization must *always* synchronize at any point where a dependence might occur, based on a static analysis of the program, whether or not the dependence actually ever occurs at runtime. Routines that modify data objects through pointers in C programs are a frequent source of this problem within many integer applications. In these programs, a compiler (and sometimes even a programmer performing hand parallelization) will typically have to assume that *any* later reads using a pointer may be dependent on the latest write of data through *any* other pointer, even if pointers only rarely point at the same data objects in memory at runtime, since compiled code must be statically guaranteed to be correct at compile time. As a result, a significant amount of thread-level parallelism can be hidden by the way the uniprocessor code is written, and will therefore be wasted as a compiler conservatively parallelizes a program.

Note that speculation and synchronization are not mutually exclusive. A program with speculative threads can still perform synchronization around uses of dependent data, but this synchronization is optional. As a result, a programmer or feedback-driven compiler can still add synchronization into a speculatively parallelized program if that helps the program execute faster. For example, adding synchronization around one or two key dependences in a speculatively parallelized program can often produce speedup by dramatically reducing the number of violations that occur. Too much synchronization, however, tends to make the speculative parallelization too conservative, stalling too frequently, and is therefore likely to be a detriment to performance.

To support speculation, one needs special coherency hardware to monitor data shared by the threads. This hardware must fulfill five basic requirements, illustrated in Fig. 3.1. The figure shows some typical data access patterns in two threads, i and $i + 1$. Figure 3.1(a) shows how data flows through these accesses when the threads are run sequentially on a normal uniprocessor. Figures 3.1(b)–3.1(e) show how the hardware must handle key situations that occur when running threads in parallel.

1. *Forward data between parallel threads.* While good thread selection can minimize the data shared among threads, typically a significant amount of sharing is required, simply because the threads are normally generated from a program in which minimizing data sharing was not a design goal. As a result, a speculative system must be able to forward shared data quickly and efficiently from an earlier thread running on one processor to a later thread running on another. Figure 3.1(b) depicts this.

2. *Detect when reads occur too early (RAW hazards).* The speculative hardware must provide a mechanism for tracking reads and writes to the shared data memory. If a data value is read by a later thread and subsequently written by an earlier thread, the hardware must notice that the read retrieved incorrect data, since a true dependence violation has

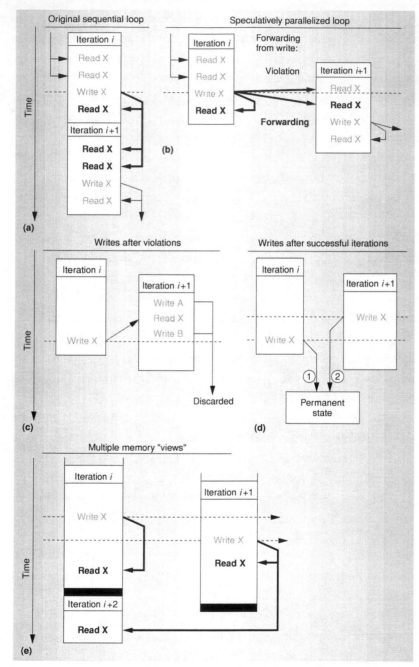

FIGURE 3.1: Five basic requirements for special coherency hardware: a sequential program that can be broken into two threads (a); forwarding and violations caused by intersections of reads and writes (b); speculative memory state eliminated following violations (c); reordering of writes into thread commit order (d); and memory renaming among threads (e).

occurred. Violation detection allows the system to determine when threads are not actually parallel, so that the violating thread can be re-executed with the correct data values. See Fig. 3.1(b).

3. *Safely discard speculative state after violations.* As depicted in Fig. 3.1(c), speculative memory must have a mechanism allowing it to be reset after a violation. All speculative changes to the machine state must be discarded after a violation, while no permanent machine state may be lost in the process.

4. *Retire speculative writes in the correct order (WAW hazards).* Once speculative threads have completed successfully, their state must be added to the permanent state of the machine in the correct program order, considering the original sequencing of the threads. This may require the hardware to delay writes from later threads that actually occur before writes from earlier threads in the sequence, as Fig. 3.1(d) illustrates.

5. *Provide memory renaming (WAR hazards).* Figure 3.1(e) depicts an earlier thread reading an address after a later processor has already written it. The speculative hardware must ensure that the older thread cannot "see" any changes made by later threads, as these would not have occurred yet in the original sequential program. This process is complicated by the fact that each processor will eventually be running newly generated threads ($i + 2$ in the figure) that will need to "see" the changes at that point in time.

In some proposed speculative hardware, the logic enforcing these requirements monitors both the processor registers and the memory hierarchy [6]. However, in systems such as Hydra, described in the next section, hardware only enforces speculative coherence on the memory system, while software handles register-level coherence by never register-allocating data that may change from thread to thread across thread boundaries.

In addition to speculative memory (or register + memory) support, any system supporting speculative threads must have a way to break up an existing program into threads and a mechanism for controlling and sequencing those threads across multiple processors at runtime. This generally consists of a combination of hardware and software that finds good places in a program to create new, speculative threads. The system then sends these threads off to be processed by the other processors in the CMP.

While in theory a program may be speculatively divided into threads in a completely arbitrary manner, in practice one is limited. Initial program counter positions and register states must be generated when threads are started, long before these would normally be known. As a result, two ways are commonly used to divide a program into threads: loops (Fig. 3.2)

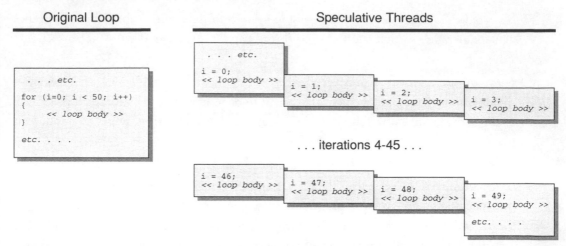

FIGURE 3.2: A graphic example of loop iteration thread-level speculation in action.

and subroutine calls (Fig. 3.3). With loops, loop-level parallelism can be exploited by starting multiple iterations of a loop body speculatively on multiple processors. As long as there are only a few straightforward loop-carried dependences, the execution of loop bodies on different processors can be overlapped to achieve speedup. Using subroutines, fine-grained task-level parallelism can be extracted using a new thread to run the code following a subroutine call's return, while the original thread actually executes the subroutine itself. As long as the return value from the subroutine is predictable (typically, when there is no return value) and any side effects of the subroutine are not used immediately, the two threads can run in parallel.

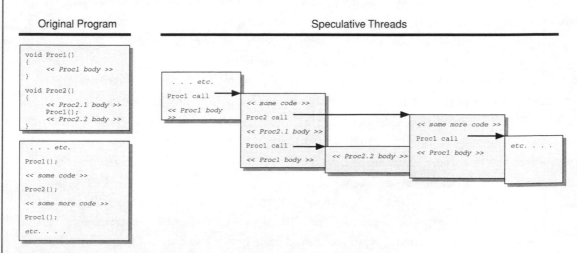

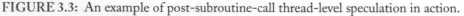

FIGURE 3.3: An example of post-subroutine-call thread-level speculation in action.

In general, achieving speedup with this technique is more challenging because thread sequencing and load balancing among the processors is more complicated with subroutines than loops.

Once threads have been created, an N-way CMP with TLS hardware must select the N least speculative threads available and allocate them to actual processors. Note that the least speculative, or "head," thread is special. This thread is actually not speculative at all, since all older threads that could have caused it to violate have already completed. As a result, it can handle events that cannot normally be handled speculatively (such as operating system calls and exceptions). Since all threads eventually become the head thread, simply stalling a thread until it becomes the head will allow the thread to process these events during speculation.

Serious consideration must be given to the size of the threads selected using these mechanisms, for the following reasons:

- *Limited buffer size.* Since one need to buffer state from a speculative region until it commits, threads need to be short enough to avoid filling up the TLS hardware buffers too often. An occasional full buffer can be handled by simply stalling the thread that has produced too much state until it becomes the "head" thread, when it may continue to execute while writing directly to memory. However, if this occurs too often, performance will suffer because opportunities for speculative execution will be greatly limited.

- *True dependences.* Excessively large threads have a higher probability of dependences with later threads, simply because they issue more loads and stores. With more true dependences, more violations and restarts occur.

- *Restart length.* A late restart on a large thread will cause much more work to be discarded on TLS systems that only take a checkpoint of the system state at the beginning of each thread. Shorter threads result in more frequent checkpoints and thus more efficient restarts.

- *Overhead.* Very small threads are also inefficient, because there is inevitably some overhead incurred during thread initiation and completion operations. Programs that are broken up into larger numbers of small threads will waste more time on these overheads.

Not all loop bodies and subroutines are of the correct size, usually on the order of thousands of instructions, but sometimes shorter, to allow TLS hardware to work efficiently. Also, many of the possible threads identifiable using these techniques have too many true dependences across loop iterations or with their calling routines to ever effectively achieve speedups during speculative execution. Since only a finite number of processors are available, care must be taken to allocate these processors to speculative threads that are likely to improve performance.

3.3 AN EXAMPLE TLS SYSTEM: HYDRA

A fairly simple implementation of TLS is implemented in the Stanford Hydra CMP. Hydra is a CMP built using four MIPS-based cores as its individual processors (see Fig. 3.4). Each core has its own pair of primary instruction and data caches, while all processors share a single, large on-chip secondary cache. The processors support normal loads and stores plus the MIPS load locked (LL) and store conditional (SC) instructions for implementing synchronization primitives.

3.3.1 The Base Hydra Design

Connecting the processors and the secondary cache together are the read and write buses, along with a small number of address and control buses. In the chip implementation, almost all buses are virtual buses. While they logically act like buses, the physical wires are divided into multiple segments using repeaters and pipeline buffers, where necessary, to avoid slowing down the core clock frequencies.

The read bus acts as a general-purpose system bus for moving data between the processors, secondary cache, and external interface to off-chip memory. It is wide enough to handle an entire cache line in one clock cycle. This is an advantage possible with an on-chip bus that all but the most expensive multichip systems cannot match due to the large number of pins that would be required on all chip packages.

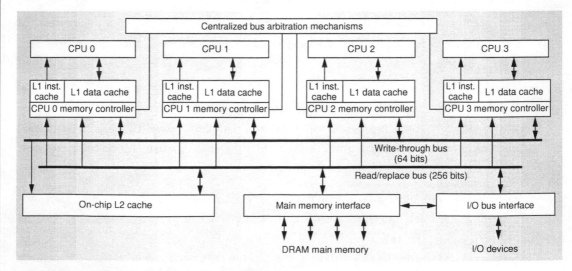

FIGURE 3.4: An overview of the Hydra CMP.

The narrower write bus is devoted to writing all writes made by the four cores directly to the secondary cache. This allows the permanent machine state to be maintained in the secondary cache. The bus is pipelined to allow single-cycle occupancy by each write, preventing it from becoming a system bottleneck. The write bus also permits Hydra to use a simple, invalidation-only coherence protocol to maintain coherent primary caches. Writes broadcast over the bus invalidate copies of the same line in primary caches of the other processors. No data is ever permanently lost due to these invalidations because the permanent machine state is always maintained in the secondary cache.

The write bus also enforces memory consistency in Hydra. Since all writes must pass over the bus to become visible to the other processors, the order in which they pass is globally acknowledged to be the order in which they update shared memory.

Hydra was designed to minimize two key design factors: the complexity of high-speed logic and the latency of interprocessor communication. Since decreasing one tends to increase the other, a CMP design must strive to find a reasonable balance. Any architecture that allows interprocessor communication between registers or the primary caches of different processors will add complex logic and long wires to paths that are critical to the cycle time of the individual processor cores. Of course, this complexity results in excellent interprocessor communication latencies—usually just one to three cycles. Past results [[ISCA 96]][7] have shown that sharing this closely is helpful, but not if it extends the access time to the registers and/or primary caches (in terms of numbers of cycles or clock rate). Consequently, register–register interconnects were omitted from Hydra. On the other hand, these earlier results also indicated that incurring the delay of an off-chip reference, which can often take 100 or more cycles in modern processors during each interprocessor communication, would be too detrimental to performance.

Because it is now possible to integrate reasonable-size secondary caches on processor dies and since these caches are typically not tightly connected to the core logic, that was the logical point of communication. In the Hydra architecture, this results in interprocessor communication latencies of 10–20 cycles, which are fast enough to minimize the performance impact from communication delays. After considering the bandwidth required by four single-issue MIPS processors sharing a secondary cache, it became clear that a simple bus architecture would be sufficient to handle the bandwidth requirements for a four- to eight-processor Hydra implementation. However, designs with more cores or faster individual processors may need to use more buses, crossbar interconnections, or a hierarchy of connections.

3.3.2 Adding TLS to Hydra

Among CMP designs, Hydra is a particularly good target for speculation because it has write-through primary caches that allow all processor cores to snoop on all writes performed. This is

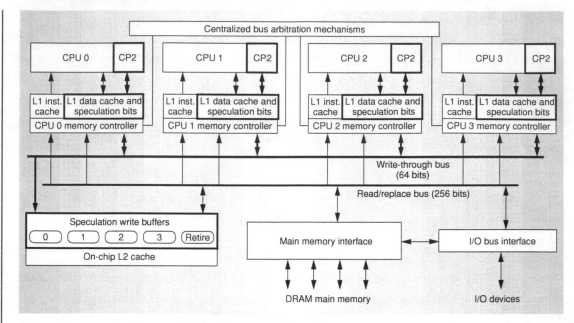

FIGURE 3.5: An overview of Hydra with speculative support.

very helpful in the design of violation-detection logic. Figure 3.5 updates Fig. 3.4 noting the necessary additions. The additional hardware is enabled or bypassed selectively by each memory reference, depending upon whether a speculative thread generates the reference.

Most of the additional hardware is contained in two major blocks. The first is a set of additional tag bits added to each primary cache line to track whether any data in the line has been speculatively read or written. The second is a set of write buffers that hold speculative writes until they can be safely committed into the secondary cache, which is guaranteed to hold only nonspeculative data. For the latter, one buffer is allocated to each speculative thread currently running on a Hydra processor, so the writes from different threads are always kept separate. Only when speculative threads complete successfully are the contents of these buffers actually written into the secondary cache and made permanent. As shown in Fig. 3.5, one or more extra buffers may be included to allow buffers to be drained into the secondary cache in parallel with speculative execution on all of the CPUs.

Other TLS architectures proposed several different mechanisms for handling speculative memory accesses produced by TLS threads. The first was the ARB, proposed along with the Multiscalar processor [8]. This was simply a data cache shared among all processors that had additional hardware to track speculative memory references within the cache. While a reasonable first concept, it requires a shared primary data cache and adds complex control logic to the data cache pipeline which has the potential to increase load latency and limit data cache

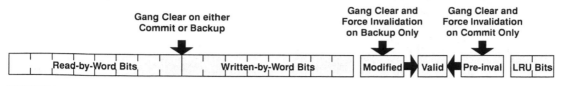

FIGURE 3.6: Additional L1 cache line bits in Hydra.

bandwidth. More recently, this was replaced with the speculative versioning cache [9], a set of separate data caches distributed among the processor cores in the Multiscalar processor that maintain their speculative state within the caches using a complex and sophisticated write-back cache protocol.

3.3.2.1 Primary Data Cache Changes

Each data cache line tag in Hydra includes several additional bits to record state necessary for speculation as shown in Fig. 3.6. The first two bits are responsible for modifying the basic cache coherence scheme that invalidates a data cache line only when a write to that line from another processor is seen on the write bus.

- *Modified bit*. This bit acts like a dirty bit in a write-back cache. If any changes are written to the line during speculation, this bit is set. These changes may come from stores by this processor or because a line is read in that includes speculative data from active secondary cache buffers. If a thread needs to be restarted on this processor, then all lines with the modified bit set are gang-invalidated at once.

- *Pre-invalidate bit*. This bit is set whenever another processor writes to the line, but is running a more speculative thread than this processor. Since writes are only propagated back to more speculative processors, a processor can safely delay invalidating the line until a different, more speculative thread is assigned to it. Thus, this bit acts as the opposite of the modified bit—it invalidates its cache line when the processor *completes* a thread. Again, all lines must be designed for gang invalidation. It must also be set if a line is loaded from the secondary cache and more speculative threads have already written to their copies of that line, since the same condition holds in that case.

The other two sets of bits allow the data cache to detect true dependence violations using the write bus mechanism. They must be designed to allow gang clearing of the bits when a speculative region is either restarted or completed.

- *Read bits*. These bits are set whenever the processor reads from a word within the cache line, unless that word's written bit is set. If a write from a less speculative thread, seen on the write bus, hits an address in a data cache with a set read bit, then a true dependence violation has occurred between the two processors. The data cache then notifies its processor with a violation exception. Subsequent stores will not activate the written bit for this line, since the potential for a violation has been established.

- *Written bits.* To prevent unnecessary violations, this bit or set of bits may be added to allow renaming of memory addresses used by multiple threads in different ways. If a processor writes to an entire word, then the written bit is set, indicating that this thread now has a locally generated version of the address. Subsequent loads will not set any read bit(s) for this section of the cache line, and therefore cannot cause violations.

Any word can have a set read or written bit, but both will never be set simultaneously. It should be noted that all read bits set during the life of a thread must be maintained until that thread becomes the head, when it no longer needs to detect dependences. Even if a cache line must be removed from the cache due to a cache conflict, the line may still cause a speculation violation. Thus, if the data cache attempts to throw out a cache line with read bits set it must instead halt the processor until the thread becomes the head or is restarted. This problem can largely be eliminated by adding a small victim buffer [10] to the data cache. This victim buffer only needs to record the address of the line and the read bits in order to prevent processor halts until the victim cache is full.

3.3.2.2 Secondary Cache Buffers
Buffering of data stored by a speculative region to memory is handled by a set of buffers added between the write bus and the secondary cache (L2). During nonspeculative execution, writes on the write bus always write their data directly into the secondary cache. During speculation, however, each processor has a secondary cache buffer assigned to it by the secondary cache buffer controller, using a simple command sent over the write bus. This buffer collects all writes made by that processor during a particular speculative thread. If the thread is restarted, then the contents of the buffer are discarded. If the thread completes successfully, then the contents are permanently written into the secondary cache. Since threads may only complete in order, the buffers therefore act as a sort of reorder buffer for memory references.

The buffers, depicted in Fig. 3.7, consist of a set of entries that can each hold a cache line of data, a line tag, and a byte-by-byte write mask for the line. As writes are made to the buffer, entries are allocated when data is written to a cache line not present in the buffer. Once a line has been allocated, data is buffered in the appropriate location and bits in the line-by-line write mask are set to show which parts of the line have been modified.

Data may be forwarded to processors more speculative than the one assigned to a particular secondary cache buffer at any time after it has been written, as is depicted in Fig. 3.8. When

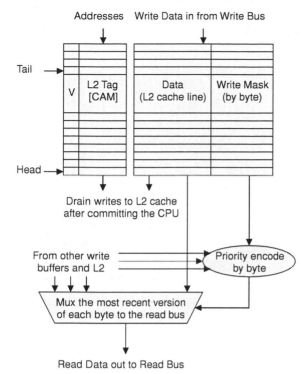

FIGURE 3.7: Speculative L2 buffers in Hydra.

one of these later processors misses in its data cache, it requests data from the secondary, as in the normal system. However, it does not just get back data from the secondary cache. Instead, it receives a line that consists of the most recent versions of all bytes in the line. This requires priority encoders on each byte to select the newest version of each byte from among this thread's buffer, all buffers from earlier threads that have not yet drained into the secondary, and the permanent value of the byte from the secondary cache itself. The composite line is assembled and returned to the requesting processor as a single, new, and up-to-date cache line. While this prioritization and byte assembly is reasonably complex, it may be done in parallel with each secondary cache read—normally a multicycle operation already.

When a buffer needs to be drained, the processor sends out a message to the secondary cache buffer controller and the procedure is initiated. Buffers drain out entry-by-entry, only writing the bytes indicated in the write mask for that entry. Since the buffers are physically located next to the secondary cache, the buffer draining may occur on cycles when the secondary cache is free, without the use of any global chip buses. In order to allow execution to continue while buffers drain into the secondary, there are more sets of buffers than processors. Whenever a processor starts a new thread, a fresh buffer is allocated to it in order to allow its previous

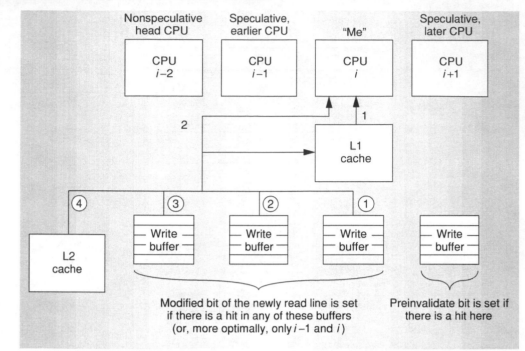

FIGURE 3.8: How secondary cache speculative buffers are read. (1) A CPU reads from its L1 cache. The L1 read bit of any hit lines are set. (2) The L2 and write buffers are checked in parallel in the event of an L1 miss. Priority encoders on each byte (indicated by the priorities 1–4 here) pull in the "newest" bytes written to a line (although no "newer" than this processor's thread). Finally, the appropriate word is delivered to the CPU and L1 with the L1 modified and preinvalidate bits set appropriately.

buffer to drain. Only in the very unlikely case when new threads are generated so quickly that all of the buffers contain data must new threads be stalled long enough to allow the oldest buffers to drain out.

Buffers may fill up during long running threads that write too much state out to memory. If these threads are not restarted, they wait until they become the head processor, write their buffers into the secondary cache, and then continue executing normally, writing directly to the secondary cache. To detect this buffer full problem, each processor maintains a local copy of the tags for the write buffer it is currently using. This local copy can detect buffer full conditions while the store that would normally overflow the buffer is executing. This store then causes an exception, much like a page fault, which allows the speculation control mechanisms to handle the situation.

3.3.2.3 Control and Sequencing

To control the thread sequencing in the Hydra system, there is a small amount of hardware added to each core using the MIPS coprocessor interface. These simple "speculation coprocessors" consist of several control registers, a set of duplicate secondary cache buffer tags, a state machine to track the current thread sequencing among the processors, and interrupt logic that can start software handlers when necessary to control thread sequencing. These software handlers are responsible for thread control and sequencing, and are summarized in Table 3.1.

As is noted in the table, some are invoked directly by software, while others act as exception handlers triggered by hardware events or messages from other processors in the system. The coprocessor maintains a table of exception vectors for speculation events, so these exception handling routines can all be started without the overhead of the operating system's normal exception dispatcher.

3.3.2.4 Putting It All Together

Together with the architecture of Hydra's existing write bus, the additional hardware allows the memory system to handle the five memory system requirements outlined previously in the following ways:

1. *Forward data between parallel threads.* When a speculative thread writes data over the write bus, all more-speculative threads that may need the data have their current copy of that cache line invalidated. This is similar to the way the system works during nonspeculative operation. If any of the threads subsequently need the new speculative data forwarded to them, they will miss in their primary cache and access the secondary cache. At this point, as is outlined in Fig. 3.8, the speculative data contained in the write buffers of the current or older threads replaces data returned from the secondary cache on a byte-by-byte basis just before the composite line is returned to the processor and primary cache. Overall, this is a relatively simple extension to the coherence mechanism used in the baseline Hydra design.

2. *Detect when reads occur too early.* Primary cache bits are set to mark any reads that may cause violations. Subsequently, if a write to that address from an earlier thread invalidates the address, a violation is detected, and the thread is restarted.

3. *Safely discard speculative state after violations.* Since all permanent machine state in Hydra is always maintained within the secondary cache, anything in the primary caches may be invalidated at any time without risking a loss of permanent state. As a result, any lines in the primary cache containing speculative data (marked with a special modified

TABLE 3.1: The speculative software handlers used by Hydra, both for speculation modes supporting fork + loop style parallelism and an optimized, loops-only variant

	ROUTINE	USE	PROCEDURE AND LOOPS OVERHEAD	LOOP-ONLY OVERHEAD	HOW USED BY SOFTWARE?
Procedures	Start Procedure	Forks off the code following a procedure call to another processor, speculatively	~70		Called by software
	End Procedure	Completes processing for a procedure that has previously forked off its completion code, when it would have returned	~110		
Loops	Start Loop	Forks off threads to execute speculative loop iterations to all processors	~75/~70	~30	
	End of each loop iteration	Completes the current loop iteration and tries to start another one	~80/16	12	
	Finish Loop	Completes the current loop iteration and shuts down loop speculation	~80	~22	

TABLE 3.1: Continued

	ROUTINE	USE	PROCEDURE AND LOOPS OVERHEAD	LOOP-ONLY OVERHEAD	HOW USED BY SOFTWARE?
Support	Violation: Local	Handles a dependency violation committed by the executing speculative thread	~30/20	7	Interrupt handler
	Violation: Receive from an "older" thread	Restarts this speculative thread after an "older" one violates	~80/11	7	
	Hold: Buffer full	Temporarily pauses a speculative thread until it becomes the "head" if the processor running this thread runs out of buffer resources	15	12	
	Hold: Exceptions	Pauses the speculative thread until it becomes the "head" following an exception	25 + OS time	17 + OS time	Exception handler

bit) may simply be invalidated all at once to clear any speculative state from a primary cache. In parallel with this operation, the secondary cache buffer for the thread may be emptied to discard any speculative data written by the thread without damaging data written by other threads or the permanent state of the machine in the secondary cache.

4. *Retire speculative writes in the correct order.* Separate secondary cache buffers are maintained for each thread. As long as these are drained into the secondary cache in the original program sequence of the threads, they will reorder speculative memory references correctly. The thread-sequencing system in Hydra also sequences the buffer draining, so the buffers can meet this requirement.

5. *Provide memory renaming.* Each processor can only read data written by itself or earlier threads when reading its own primary cache or the secondary cache buffers. Writes from later threads do not cause immediate invalidations in the primary cache, since these writes should not be "visible" to earlier threads. This allows each primary cache to have its own local copy of a particular line. However, these "ignored" invalidations are recorded using an additional pre-invalidate primary cache bit associated with each line. This is because they must be processed before a different speculative or nonspeculative thread executes on this processor. If a thread has to load a cache line from the secondary cache, the line it recovers only contains data that it should actually be able to "see," from its own and earlier buffers, as Fig. 3.8 indicates. Finally, if "future" threads have written to a line in the advancing processor's primary cache, the pre-invalidate bit for that line will be set, either during a snoop on the write when it occurred or when the line was later loaded from the secondary cache and buffers. When the current thread completes, these bits allow the processor to quickly simulate the effect of all stored invalidations caused by all writes from later processors all at once, before a new thread begins execution on this processor.

Based on the amount of memory and logic required, the cost of adding speculation hardware should be comparable to adding an additional pair of primary caches to the system. This enlarges the Hydra die only by a few percent.

3.3.3   Using Feedback from Violation Statistics

A key advantage of a TLS system over a conventional parallel system is that it facilitates profile-directed optimization of parallel code. In a real system, this feedback could be obtained by adding speculative load program counter memory to each processor, broadcasting store program counters along with data on the write bus, and then combining the results from these hardware structures using instrumentation code built into the speculation software routines (at profiling time only—normally the overhead imposed by such code could be "turned off") that would match up load/store pairs causing violations and record them for later analysis. Combined with information on the amount of time lost to each violation and to stalling, this information could then be fed into a profiling compiler framework or simply presented to the user through a simple interface to speed TLS code using one or more techniques.

3.3.3.1 Explicit Synchronization

Simply by adding a way to issue a nonspeculative load instruction even while the processor is executing speculatively, one can add explicit synchronization support to speculative hardware. As depicted in Fig. 3.9, this special load may be used to test lock variables that protect the critical regions of code in which pairs of loads and stores exist that cause frequent dependence violations. Before entering a critical region, synchronizing code spins on the lock variable until the lock is released by another processor. Because the special load is nonspeculative, a violation does not occur when the lock is released by a store from another processor. Once the lock is freed, the speculative processor may perform the load at the beginning of the critical region. Finally, when a processor has performed the store at the end of the region, it updates the lock so that the next processor may enter the critical region. This process eliminates all restarts caused by dependent load–store pairs in the critical region, at the expense of spinlock stalls and forcing the speculative processors to serialize through the critical regions, eliminating any possibility of finding parallelism there (which was presumably an impossible task, anyway). The lock handling code also adds a small software overhead to the program.

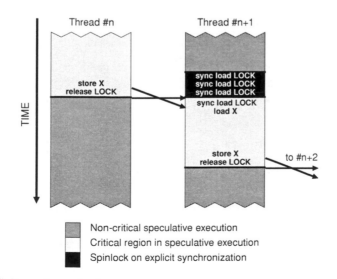

FIGURE 3.9: Explicit synchoronization.

3.3.3.2 Code Motion

While explicit synchronization prevents critical dependences from causing violations, it also forces the speculative processors to serialize their execution. For small critical regions, this is perfectly acceptable, but for large ones it can easily eliminate all of the parallelism that speculative threads are attempting to exploit. To avoid this situation, the preferred technique to improve speculation performance is to move dependent loads and stores in order to shrink the critical regions between frequently violating load–store pairs. This makes it possible to reduce the number of violations and often increases the inherent parallelism in the program by lengthening the sections of code that can be overlapped on different processors without causing violations.

This works in two ways. At the tops of critical regions, loads can sometimes be delayed by rearranging code blocks in order to move code without critical dependences higher up in the loop body. However, this is usually only possible in large loops built up from several nondependent sections of code that can be interchanged freely. Therefore, it is more common to make stores to shared variables occur earlier. Induction variables are an obvious target for early stores. Since the store that updates the induction variable is not dependent upon any computation within the loop, updates to these variables can safely be moved to the top of the loop body. Other variables, that do depend upon results calculated in a loop iteration, will not be improved as dramatically by scheduling their critical stores early, but performance can often still be improved significantly over unmodified code using these techniques. It should also be noted that code motion on a speculative processor is somewhat different from that on a conventional multiprocessor, since *only* the most critical loads and stores of variables need to be moved to reduce the potential for restarts. For example, variables that often—but not always—change in a predictable way, such as induction variables, can be *speculatively* precomputed and updated at the top of the loop. Should they later be updated in a less predictable manner, the variable may simply be rewritten, probably causing restarts on the more speculative processors. However, as long as the value predicted and written early is used most of the time, the amount of available parallelism in the program may be vastly increased.

3.3.3.3 Parallel Reduction Optimization

Parallel reduction transformations allow certain iterative functions to be parallelized. For example, iterative accumulations into a single summation variable could be transformed into four parallel summations that are only combined at the end of the loop. In general, these optimizations can be performed with any sort of associative accumulation (using addition, multiplication, logical OR, logical AND, etc.), and can greatly increase the number of loops that are good candidates for parallelization using TLS. These techniques have been used for years with conventional parallelizing compilers [11].

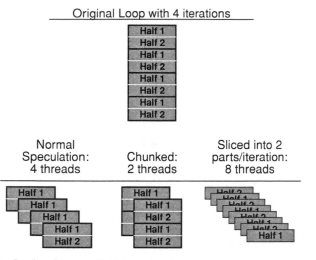

FIGURE 3.10: Loop body chunking and slicing.

3.3.3.4 Loop Body Slicing and Chunking

More radical forms of code motion and thread creation are possible by breaking up a loop body into smaller chunks that execute in parallel or the converse process, combining multiple speculative threads together into a single thread. With the former technique, loop slicing, a single loop body is spread across several speculative iterations running on different processors, instead of running only as a single iteration on a single processor. In the latter case, loop chunking, multiple loop iterations may be chunked together into a single, large loop body. Loop bodies that were executed on several processors are combined and run on one. This generally only results in better performance if there are no loop-carried dependences besides induction variables, which limits the versatility of loop chunking. However, if a completely parallel loop can be found, chunking can allow one to create speculative threads of nearly optimal size for the speculative system. Figure 3.10 shows conceptually how slicing and chunking work.

While loop chunking only needs to be performed if the body of the loop is so small that the execution time is dominated by speculation overheads, the motivations for loop slicing are more complex. The primary reason is to break down a single, large loop body, made up of several fairly independent sections, into smaller parts if they are more optimally sized for speculation. As was mentioned previously, in a very large loop body a single memory dependence violation near the end of the loop can result in a large amount of work being discarded. Also, the large loop body may overflow the buffers holding the speculative state. Buffer overflow prevents a speculative processor from making forward progress until it becomes the head, nonspeculative processor, so this should normally be avoided whenever possible. Loop slicing is also a way to

```
for (x=0; x<1000; x++)
    if ((x % 10) == 0)
        for (y=0; y<10; y++)
            InnerLoopOneThousandCyclesOfWork();
    OuterLoopOneThousandCyclesOfWork();
```

FIGURE 3.11: Code with a two-level loop structure.

perform code motion to prevent violations. If there is code in the loop body that calculates values that will be used in the next loop iteration and this code is not usually dependent upon values calculated earlier in the same loop iteration, then this code may be sliced off of the end of the loop body and assigned to its own speculative thread. In this way, the values calculated in the sliced-off region are essentially "precomputed" for later iterations, since they are produced in parallel with the beginning of the loop iteration. The advantage of slicing over normal code motion within a single thread is that no data dependence analysis is required to ensure that it is legal to perform the code motion, since the violation detection mechanism will still enforce all true dependences that may exist. Another way in which a version of loop slicing is useful is depicted in Fig. 3.11, which shows how parallelism can exist at multiple levels within a set of nested loops, making parallelization of either loop alone suboptimal. Assuming that each of the thousand-cycle routines is a fairly independent task, if only either the outer loop or the inner loop is parallelized, half the TLP that exists will not be extracted.

Loop chunking may be implemented in a compiler in a similar manner to the way loop unrolling is implemented today, since both are variations on the idea of combining loop iterations together. In fact, the two operations may be merged together, so that a loop is unrolled as it is chunked into a speculative thread. As a result, adding this capability to current compilers should not be difficult. Effective slicing, however, requires more knowledge about the execution of a program, although the ability to analyze the control structure of the application combined with some violation statistics should be sufficient. Slicing should not be performed indiscriminately because it may allow speculative overheads to become significant and/or it may result in significant load imbalance among the processors if some slices are much larger than others.

3.3.4 Performance Analysis

The Hydra system with speculation support (as described in Table 3.2) was simulated using a variety of C language benchmarks. Table 3.3 lists the selection of applications chosen. Many of these programs are difficult or impossible to parallelize using conventional means due to the presence of frequent true dependences. Automatically parallelizing compilers are stymied

TABLE 3.2: The system configuration of Hydra used for simulations

CHARACTERISTIC	L1 CACHE	L2 CACHE	MAIN MEMORY
Configuration	Separate I and D SRAM cache pairs for each CPU	Shared, on-chip SRAM cache	Off-chip DRAM
Capacity	16 Kbytes each	2 Mbytes	128 Mbytes
Bus width	32-bit connection to CPU	256-bit read bus + 32-bit write bus	64-bit bus at half CPU speed
Access time Associativity	1 CPU cycle 4 way	5 CPU cycles 4 way	At least 50 cycles N/A
Line size	32 bytes	64 bytes	4-Kbyte pages
Write policy	Write through, no allocate on write	Write back, allocate on writes	Write back (virtual memory)
Inclusion	N/A	Inclusion enforced by L2 on L1 caches	Includes all cached data

by the presence of many C pointers in the original source code that they cannot statically disambiguate at compile time. On all applications except *eqntott*—which was parallelized using subroutine speculation—loops in the original programs were just converted to their speculative forms.

Figure 3.12 summarizes the results. After initial speculative runs with unmodified loops from the original programs, feedback was used to optimize the benchmarks' source code by hand. This avoided the most critical violations that caused large amounts of work to be discarded during restarts. These optimizations were usually minor—usually just moving a line of code or two or adding one synchronization point [12]. Notably, the modifications required were always orders of magnitude simpler than those required to change these applications for use with conventional parallelization, where that was even possible. However, they had a dramatic impact on benchmarks such as *MPEG2*.

Overall, these results are at least comparable to and sometimes better than a single large uniprocessor of similar area running these applications, based on the results presented in the study of Chapter 1 [13].

TABLE 3.3: A summary of the speculatively parallelized applications used to make performance measurements with Hydra. Applications in italics were also handparallelized and run on the base Hydra design

APPLICATION	SOURCE	DESCRIPTION	HOW PARALLELIZED
compress	SPEC95	Entropy-encoding compression of a file	Speculation on loop for processing each input character
eqntott	SPEC92	Logic minimization	Subroutine speculation on core quick sort routine
grep	Unix command	Finds matches to a regular expression in a file	Speculation on loop for processing each input line
m88ksim	SPEC95	CPU simulation of Motorola 88000	Speculation on loop for processing each instruction
wc	Unix command	Counts the number of characters, words, and lines in a file	Speculation on loop for processing each input character
ijpeg	SPEC95	Compression of an RGB image to a JPEG file	Speculation on several different loops used to process the image
MPEG2	Mediabench suite	Decompression of an MPEG-2 bistream	Speculation on loop for processing slices
alvin	SPEC92	Neural network training	Speculation on 4 key loops
cholesky	Numeric recipes	Cholesky decomposition and substitution	Speculation on main decomposition and substitution loops
ear	SPEC92	Inner ear modeling	Speculation on outer loop of model
simplex	Numeric recipes	Linear algebra kernels	Speculation on several small bitstream loops

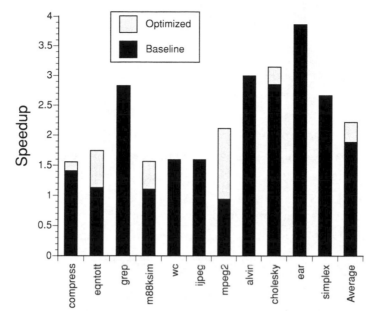

FIGURE 3.12: Speedup of speculatively parallelized applications running on Hydra compared with the original uniprocessor code running on one of Hydra's four processors. The gray areas show the improved performance following tuning with feedback-based code.

Of course, a CMP can also perform faster by running fully parallelized programs without speculation, when those programs are available. A uniprocessor cannot. It is even possible to mix and match using multiprogramming. For example, two processors could be working together on a speculative application, while others work on a pair of completely different jobs. While it was not implemented in the Hydra prototype, one could relatively easily enhance the speculative support routines so that multiple speculative tasks could run simultaneously. Two processors could run one speculative program, and two could run a completely different speculative program. Since additional speculative processors provide diminishing performance benefits, this would allow the overall system performance to be maximized through a combination of nonspeculative and speculative threading. In this manner, it is possible for a CMP to nearly always outperform a large uniprocessor of comparable area.

Speedups are only a part of the story, however. Speculation also makes parallelization much easier, because a parallelized program that is guaranteed to work exactly like the uniprocessor version can be generated automatically. As a result, programmers only need to worry about choosing which program sections should be speculatively parallelized and then doing some tweaks for performance optimization. Even when optimization is required, speculative parallelization typically took a single programmer a day or two per application. In contrast, hand parallelization of these C benchmarks typically took one programmer anywhere from a week to a month, since it was necessary to worry about correctness *and* performance throughout the process. As a result, even though adding speculative hardware to Hydra makes the chip somewhat harder to design and verify, the reduced cost of generating parallel code offers significant advantages.

3.3.5   Completely Automated TLS Support: The JRPM System

Jrpm is a dynamic extension to the Hydra TLS system that allows *completely automatic* parallelization of general Java-language programs. A similar, albeit more complex, scheme could be used for programs written in languages that are statically compiled down to assembly language, such as C. As a proof-of-concept, however, it was much simpler to work with a dynamically-compiled language such as Java. Jrpm parallelizes programs with almost no input from the user or programmer. Its custom runtime system with special hardware support analyzes dynamic execution for parallelism and correctly handles dynamic dependences. Figure 3.13 shows the system's key components:

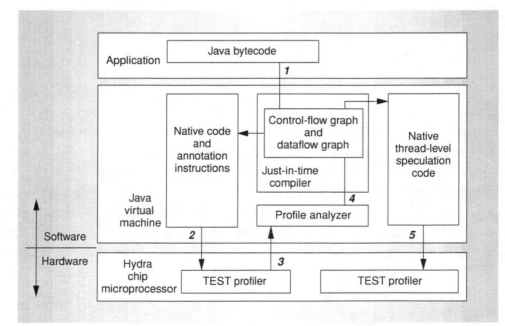

FIGURE 3.13: Overview of the JRPM system, including hardware and software components. Programs running on JRPM execute the following steps: 1. Identify thread decompositions by analyzing bytecodes, and compile natively with annotation instructions. 2. Run annotated program sequentially, collecting TEST (Tracer for Extracting Speculative Threads) profile statistics on potential thread decompositions. 3. Postprocess profile statistics and choose thread decompositions that provide the best speedups. 4. Recompile code with thread-level speculation (TLS) instructions for selected thread decompositions. 5. Run native TLS code.

- *Hardware profiler.* Static parallelizing compilers have insufficient information to analyze dynamic dependences effectively. Dynamic analysis to find parallelism complements a TLS processor's ability to parallelize optimistically and to use hardware to guarantee correctness. TEST (Tracer for Extracting Speculative Threads) support analyzes sequential program execution in real time to find the best regions to parallelize with minimum hardware support.

- *Software virtual machine.* Virtual machines such as Sun's JVM and Microsoft's .NET VM have become commercially popular for supporting platform-independent applications. In Jrpm, the JVM acts as an abstraction layer that hides the dynamic analysis framework and thread-level speculation from the program, letting us seamlessly support a new execution model without modifying the source binaries.

- *Java application.* Written in bytecodes that are portable from one system to another and can be recompiled by the JVM into native code for each machine.

Following Fig. 3.13, the compiler derives a control flow graph (CFG) from program bytecodes and analyzes it to identify potential thread decompositions [14]. A single Hydra processor executes, as a sequential program, a Java program that has been dynamically compiled with instructions annotating local variables and possible thread decompositions. Trace hardware collects statistics in real time for the prospective decompositions. Once this hardware has collected sufficient data, the dynamic compiler recompiles into speculative threads those regions predicted to have the largest speedup and that will exploit as much parallelism as possible from the program.

Although the primary goal of the Jrpm dynamic parallelization system is to automatically speed up program execution, the system also benefits from additional properties that are attractive to both programmers and system designers:

- *Reduced programmer effort.* Manually identifying fine-grained parallel decompositions can be time consuming, especially for programs without obvious critical sections. Because Jrpm automatically selects and guarantees the correct behavior of executing parallel threads, programmers can focus on performance debugging instead of the usual complexities of parallel programming.

- *Portability.* Jrpm works with unmodified sequential-program bytecodes. Because the system doesn't modify the binaries explicitly for TLS, the code retains its platform independence.

- *Retargetability.* Because parallel decompositions are not explicitly coded, Jrpm can dynamically adapt decompositions at runtime for future chip multiprocessors with more processors, larger speculative buffers, or different cache configurations.

- *Simplified analysis.* Compared to traditional parallelizing compilers, the Jrpm system relies on more hardware for TLS and profiling support, but reduces the complexity of the analysis required to extract exposed thread-level parallelism from both floating-point and difficult-to-analyze integer applications.

3.3.5.1 Tracer for Extracting Speculative Threads

TLS simplifies many automatic parallelization challenges, but any sort of automated paralleliz-ing tool has to consider certain constraints when selecting regions for this execution model. To review, the major constraints are as follows.

- True interthread data dependences, or read-after-write hazards, always limit speedup from parallel execution of speculative threads.

- Speculative read and write states buffered by the hardware cannot be discarded during speculative execution and must fit into the on-chip hardware structures. Attempts to drop an L1 cache line with speculative read bits set or to write to a full store buffer will cause a stall until the thread becomes the nonspeculative head thread and safe execution of loads or stores is possible.

- Only one thread decomposition (for example, one loop in a loop nest) can be active at a time.

- Compiled speculative thread code introduces sequential overheads from speculative thread management routines and forced communication of interthread dependent local variables, limiting speedups under TLS for very small threads (say, with less than 10 instructions) [15, 16].

Dynamic analysis to identify appropriate speculative thread loops (STLs) complements the baseline Hydra-with-TLS processor's ability to parallelize optimistically and to use hardware to guarantee correctness. This analysis is difficult because the constraints impose conflicting requirements for selecting thread decompositions. Speculating on small loops limits parallel coverage and suffers from higher speculative-thread overheads relative to the work performed. Speculating on large loops increases the probability of speculation buffer overflows and could incur higher relative dependence-violation penalties. The automated parallelizing tool must resolve these conflicts using a set of heuristic rules.

3.3.5.2 Analysis Overview

The compiler examines a method's CFG to identify all natural loops that could be a potential STL [11]. Two types of trace analyses characterize an STL's potential: load dependences and speculative state overflow.

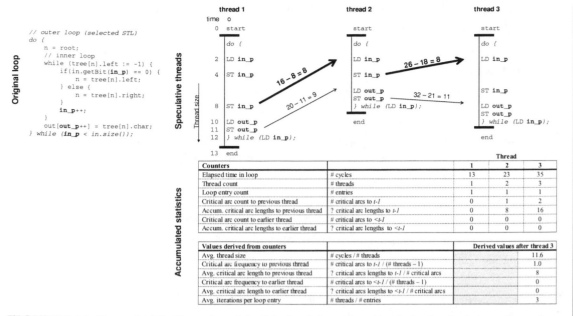

Original loop

```
// outer loop (selected STL)
do {
    n = root;
    // inner loop
    while (tree[n].left != -1) {
        if(in.getBit(in_p) == 0) {
            n = tree[n].left;
        } else {
            n = tree[n].right;
        }
        in_p++;
    }
    out[out_p++] = tree[n].char;
} while (in_p < in.size());
```

Counters		1	2	3
Elapsed time in loop	# cycles	13	23	35
Thread count	# threads	1	2	3
Loop entry count	# entries	1	1	1
Critical arc count to previous thread	# critical arcs to *t-1*	0	1	2
Accum. critical arc lengths to previous thread	? critical arc lengths to *t-1*	0	8	16
Critical arc count to earlier thread	# critical arcs to <*t-1*	0	0	0
Accum. critical arc lengths to earlier thread	? critical arc lengths to <*t-1*	0	0	0

Values derived from counters		Derived values after thread 3
Avg. thread size	# cycles / # threads	11.6
Critical arc frequency to previous thread	# critical arcs to *t-1* / (# threads – 1)	1.0
Avg. critical arc length to previous thread	? critical arcs lengths to *t-1* / # critical arcs	8
Critical arc frequency to earlier thread	# critical arcs to <*t-1* / (# threads – 1)	0
Avg. critical arc length to earlier thread	? critical arcs lengths to <*t-1* / # critical arcs	0
Avg. iterations per loop entry	# threads / # entries	3

FIGURE 3.14: Example (Huffman decode) of the load dependence analysis. Analysis is performed on the outer loop in this example. Loop-carried dependences are bold in source code. Arrows represent dependence arcs. Critical arcs are shown in darker arrows.

By examining executing loads and stores, load dependence analysis looks for interthread dependences for an STL. TEST records the time stamp when a memory or local-variable store occurs; on subsequent loads to the same address, TEST retrieves this time stamp. By comparing this value with the thread-start time stamp, it is possible to detect the frequency of interthread dependence arcs and identify critical arcs. (A critical arc is the shortest dependence arc that limits parallelism between a given pair of threads.) An example of this analysis on a Huffmann decoder core loop is shown in Fig. 3.14.

Speculative-state-overflow analysis checks that the speculative state generated by an iteration of an STL will fit within the limits of the L1 caches and store buffers. TEST maintains a history of cache lines accessed by loads and stores. From this, TEST can determine the approximate speculative memory footprint of the current speculative thread. By maintaining counters tracing these requirements, TEST can estimate how frequently a given STL will overflow its speculative buffer limits.

Once TEST has collected enough profiling data (for example, at least thousands of iterations of an STL under analysis), it computes the estimated speedup for each STL from the dependence arc frequencies, thread sizes, critical arc lengths, overflow frequencies, and speculative overheads. Using statistics from the two analyses and the computed speedup, Jrpm recompiles into speculative threads only those loops that have many average loop iterations per entry, seldom overflow speculative buffers, a predicted speedup greater than 1.2, and are executed more than 0.5% of the time during sequential execution. It is often possible to choose multiple decompositions in a loop nest. In this case, Jrpm selects the best STL by comparing the estimated execution time for the different STL decompositions provided by each nesting.

3.3.5.3 Hardware–Software Support for TEST

The hardware to minimize profiling overheads and improve accuracy analyzes a sequentially executing program and therefore works only when speculation is disabled. Hence, TEST can reuse some of the speculative hardware for profiling purposes, since it would otherwise be idle during profiling.

Annotation instructions that the dynamic compiler inserts into native code mark important events relevant to trace analyses. Annotations mark a potential STL's entry, exit, and iteration end. TEST uses explicit annotations to track local variables in the same calling context as a potential STL and that could cause dependences. This simplifies the tracking of these variables in optimized compiled code. A processor automatically communicates memory load and store events to the tracing hardware when tracing is enabled. At the end of an STL (for example, an exit from a loop), special routines read the collected statistics from TEST for use by the runtime system.

The annotation instructions communicate events to small banks of hardware comparators that carry out the bulk of the dependence and overflow trace analyses. Each comparator bank, built using a small number of comparators and counters, tracks the progress for a given STL by analyzing and collecting statistics on incoming loads and stores. Having an array of comparator banks allows the tracing of multiple potential STLs that execute concurrently, such as different levels in loop nests. Calculations suggest that an implementation of the TEST hardware with eight comparator banks would add less than 1% to the transistor count of the Hydra chip multiprocessor with TLS support.

The speculative store buffers, which are normally idle during sequential nonspeculative execution, hold a history of previous time stamp events during profiling. The buffers retrieve an address' time stamp on an annotating memory or local variable instruction for use in the comparator banks. The store buffers, organized as first-in, first-out (FIFO) buffers during tracing, effectively hold a limited history of memory and local-variable accesses.

3.3.5.4 Compiling Selected Regions into Speculative Threads

Jrpm's Java runtime system is based on the open-source Kaffe virtual machine (http://kaffe.org) [17], but a custom just-in-time compiler, microJIT, and a garbage collector were added to make up for the original virtual machine's performance limitations. The microJIT compiler was augmented to generate speculative thread code. The dynamic compiler inserts speculative-thread-control routines into the STLs chosen by TEST analysis. In addition to the fixed speculative-handler overheads, additional overheads are possible in certain circumstances. The master processor must communicate STL initialization values to the slave processors by saving them to the runtime stack. Certain optimizations must insert cleanup code at the entry and exit of STLs. Furthermore, the compiler must force local variables that could cause interthread (loop-carried) dependences in an STL to communicate through loads and stores in a runtime stack shared between all speculative processors.

When possible, Jrpm's dynamic compiler automatically applies optimizations to improve speculative performance for selected STLs. Table 3.4 summarizes these compiler optimizations.

3.3.5.5 Parallelizing Real Programs Using Jrpm

Table 3.5 summarizes the characteristics of the STLs automatically chosen from TEST analysis. Overall, there was significant diversity in the parallel coverage of selected STLs. Although many programs have single critical sections, *Assignment, NeuralNet, euler,* and *mp3* have many STLs that contribute equally to total execution time. Several programs have more selected STLs than those shown in the table, but the omitted decompositions do not have any significant coverage. The *mp3, db, jess,* and *DeltaBlue* benchmarks have significant sections of serial execution that are not covered by any potential STLs, limiting the total speedup for these applications. These benchmarks come from the jBYTEmark [18], SPECjvm98 [19], and Java Grande [20] suites, as well as real applications found on the Internet.

TLS can simplify program parallelization, but not all programs can benefit from it. Some integer benchmarks evaluated using TEST show no potential for speedup using speculation. Programs with system calls in critical code loops do not speed up on Jrpm, because the Hydra implementation of TLS cannot handle system calls speculatively. Several other integer programs contain only loops that consistently overflowed the speculative state, executed too few iterations for speculation to be effective, or contained an unoptimizable serializing dependence.

The larger programs contain so many loops that manual identification of STLs would have been too time consuming. A visual analysis of the source code revealed that a traditional parallelizing compiler could analyze less than half the benchmarks.

3.3.5.6 Performance Results

Each benchmark was run as a sequential annotated program on Jrpm with the TEST profiling system enabled. The dynamic compiler then recompiled the benchmark and executed it using speculative threads with the STLs selected by TEST. Figure 3.15 shows slowdown during profiling, the predicted TLS execution time from TEST analysis, and actual TLS performance. Figure 3.16 compares total program speedup (adding compilation, garbage

TABLE 3.4: Summary of low-level TLS compiler optimizations used by JRPM

OPTIMIZATION	FUNCTION	BENEFIT	COST
Loop-invariant register allocates	Register allocation memory load that always returns the same value	Eliminates redundant memory load per iteration	Load of value into the register at init and restart
Noncommunicating loop inductor	Locally computes loop inductor value for a thread	Eliminates frequent RAW violations for loop inductors incremented at end of iteration	Computation of inductor value at init and restart
Resetable loop inductor	Locally computes loop inductor-like value for a thread	Eliminates frequent RAW violations for loop inductor-like values	Computation of loop inductor-like value at init and restart
Reduction	Computes associative operations locally	Eliminates dependencies for associative operations	Merge locally computed values for final reduction value at exit
Synchronizing lock	Protects loop-carried dependencies from spurious RAW violations	Eliminates RAW violations from frequent dependencies	Wait and signal overhead of the lock for every thread
Multilevel decompositions	Switches selected STLs between an outer and inner loop in a nested loop	Improves load balancing for irregularly structured nested loops	Init and exit overhead for switching between STL decompositions

TABLE 3.5: Description and characteristics of integer benchmarks evaluated on the JRPM system

CATEGORY	BENCHMARK	DESCRIPTION	LOOP COUNT	NO. OF SELECTED LOOPS (STLS)	NO. OF LOOP ITERATIONS (SPECULATIVE THREADS) PER STL	LENGTH OF THREADS ON A SINGLE-ISSUE PROCESS, IN CYCLES	% OF SERIAL EXECUTION AFTER PARALLELIZING
Integer	Assignment	Resource allocation	32	11	29	199	1%
	BitOps	Bit array operations	4	2	7.646	29	0%
	compress	Compression	28	4	93,755	546	0%
	db	Database	37	6	23,142	510	3%
	deltaBlue	Constraint solver	22	5	82	501	22%
	EmFloatPnt	Floating-point emulation	7	1	255	20,127	0%
	Huffman	Compression	14	6	502	108	0%
	IDEA	Encryption	2	1	242	6,307	0%
	jess	Expert system	134	4	166	339	27%
	JLex	Lexical analyzer generator	128	7	71	2,699	7%
	MipsSimulator	CPU simulator	19	2	51,931	1,313	0%
	monteCarlo	Monte Carlo simulation	15	5	942	119	5%
	NumHeapSort	Heap sort	5	2	6,081	555	0%
	raytrace	Ray tracer	14	1	65	158	9%
Floating Point	euler	Fluid dynamics	32	13	66	304	1%
	fft	Fast Fourier transform	5	2	187	231	0%
	FourierTest	Fourier coefficients	2	1	100	167,802	0%
	LuFactor	LU factorization	13	7	64	455	0%
	moldyn	Molecular dynamics	8	1	1,026	96	2%
	NeuralNet	Neural net	19	8	9	617	1%
	shallow	Shallow water simulation	11	3	257	1,420	0%
Multimedia	decJpeg	Image decoder	61	21	34	124	13%
	encJpeg	Image compression	62	9	54	121	1%
	h263dec	Video decoder	54	3	165	212	10%
	mpegVideo	Video decoder	69	9	23	701	47%
	mp3	Mp3 decoder	98	17	55	181	14%

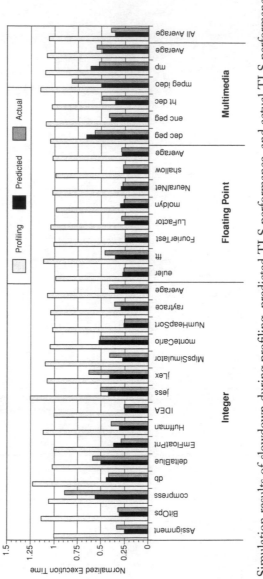

FIGURE 3.15: Simulation results of slowdown during profiling, predicted TLS performance, and actual TLS performance on Hydra, normalized to the original sequential program's execution time.

FIGURE 3.16: Total program speedup with compilation, garbage collection, profiling, and recompilation overheads using default benchmark data sets.

collection, profiling, and recompilation overheads) normalized with respect to normal serial execution (including compilation and garbage collection overheads) for a given benchmark run.

During profiling, most benchmarks experience no more than a 10% slowdown, and only two applications have slowdowns approaching 25%, as Fig. 3.15 shows. These slowdowns are reasonable, especially considering the relatively short period of time that most programs must spend on profiling to select an STL.

Simulations of this system show that the Jrpm approach has significant potential for automatically exploiting thread-level parallelism. From the wide set of Java benchmarks, thread parallelism could be exploited from integer, floating-point, and multimedia benchmarks. The best speedups, approaching 4×, occur with the floating-point applications. The speedups achieved on multimedia and integer programs are also significant, between 1.5× and 3×, but vary widely and are generally less than those achieved for floating-point applications.

Overall, TLS execution characteristics such as average thread size and number of threads per loop entry (see Table 3.5) vary widely from program to program. Despite this, the average thread size for most benchmarks is at least 100 instructions (also cycles on the single-issue Hydra processors). The average thread size appears large enough to suggest that programs could benefit further from superscalar cores that exploit instruction-level parallelism relatively independent of the coarse-grained parallelism that TLS targets.

The overheads for profiling and dynamic recompilation are small, even for the shorter-running benchmarks. Contributing factors include the low-overhead profiling system, the limited profiling information required to make reliable STL choices, and the small amount of code that must be recompiled to transform a loop. In these benchmarks, selected STLs vary little with the amount of profiling information collected, once TEST collects enough data to overcome local variations in RAW violations, buffer overflows, and thread sizes. The reason for this stability is that most selected STLs are invariant to the input data set. For benchmarks with STLs sensitive to the input data set, the input data sets remain stable for the duration of the benchmark. In real-world cases, in which the input data sets can change during runtime, Jrpm could trigger reprofiling and recompilation when a selected STL sensitive to the input data set consistently experiences unexpected behavior, in the form of excessive numbers of dependence violations.

For six of the integer benchmarks, some programmer assistance is needed to expose sufficient parallelism. The transformations were achieved with the assistance of TEST feedback. TEST profiling results that summarized critical potential STLs and associated dependence arc frequencies and lengths facilitated quick identification of performance bottlenecks in the source code. The resulting transformations, listed in Table 3.6, significantly improve performance and do not slow down the original sequential execution. Only three of these benchmarks require significant manual transforms, while the other three need only trivial modifications. Examination of the program sources suggests that most of these modifications cannot be performed automatically because they require high-level understanding of the program.

TABLE 3.6: Difficulty and potential for compiler automation of manual transformations performed that improve speculative performance

BENCHMARK	DIFFICULTY	COMPILER OPTIMIZABLE	LINES MODIFIED	MODIFIED OPERATIONS
NumHeapSort	Low	N	7	Remove loop carried dependency at top of sorted heap
Huffman	Med	N	22	Merge independent streams to prevent sub-word dependencies during compression. Guess next offset when uncompressing data stream
MipsSimulator	Med	N	70	Minimize dependencies for forwarding load delay slot value
db	Low	Y	4	Schedule loop carried dependency
compress	Low	N	13	Guess next offset when compressing/uncompressing data
monteCarlo	Med	N	39	Schedule loop carried dependency

3.4 CONCLUDING THOUGHTS ON AUTOMATED PARALLELIZATION

All of these techniques are helpful, because they provide a way to use multiple cores in a CMP with code that has been written for traditional uniprocessor programming models, but they are far from optimal. From a programmer's point of view, all attempt to keep an illusion of a conventional uniprocessor intact, even though the work will actually be performed across many processor cores within a CMP. This is a *great* advantage, because it allows existing programs to be ported to CMPs with virtually no effort. However, all of these techniques involve using the additional thread(s) to perform essentially speculative operations, such as L2 prefetches or actual speculative execution, in an effort to accelerate the "main" thread. Like all speculative execution, there is a great potential for triggering large quantities of wasted work on any sort of program that executes in rather unpredictable ways. All of this wasted work, naturally, translates into wasted processor resources and energy, which severely undercuts the advantages that a CMP can ordinarily offer in these areas over a conventional uniprocessor.

As a result, if it is possible to update your programs into *truly* parallel applications, with multiple more-or-less independent threads, then this is generally the best way to utilize a CMP in order to avoid wasting effort on speculative work. Of course, even if original source code and programmer time is available, converting applications to use multiple explicit threads is a very hard task with today's parallel programming models, so much so that only a tiny fraction of programmers ever attempt the effort required. While the short interprocessor communication latencies in a CMP make the task of parallelization somewhat easier by minimizing the effects of poor interthread communication planning, issues such as deadlock avoidance, locking design, and the like must be carefully considered and evaluated. In the next chapter we will examine features that can be added to CMPs to ease these problems.

REFERENCES

[1] R. S. Chappell, F. Tseng, A. Yoaz, and Y. N. Patt, "Difficult-path branch prediction using subordinate microthreads," in *Proc. 29th Annual Int. Symp. Computer Architecture*, Anchorage, AK, June 2002, pp. 307–317.

[2] D. Kim, S. S. Liao, P. H. Wang, J. del Cuvillo, X. Tian, X. Zou, H. Wang, D. Yeung, M. Girkar, and J. P. Shen, "Physical experimentation with prefetching helper threads on Intel's hyper-threaded processors," in *Proc. Int. Symp. Code Generation and Optimization (CGO 2004)*, Palo Alto, CA, Mar. 2004, pp. 27–38.

[3] Y. Song, S. Kalogeropulos, and P. Tirumalai, "Design and implementation of a compiler framework for helper threading on multi-core processors," in *Proc. 14th Int. Conf. on Parallel Architectures and Compilation Techniques (PACT-2005)*, St. Louis, MO, Sept. 2005, pp. 99–109.

[4] C.-K. Luk, "Tolerating memory latency through software-controlled pre-execution in simultaneous multithreading processors," in *Proc. 28th Annual Int. Symp. Computer Architecture (ISCA-28)*, Göteborg, Sweden, June 2001, pp. 40–51.

[5] Standard Performance Evaluation Corporation, SPEC, http://www.spec.org, Warrenton, VA.

[6] G. Sohi, S. Breach, and T. Vijaykumar, "Multiscalar processors," in *Proc. 22nd Annual Int. Symp. Computer Architecture*, Santa Margherita Ligure, Italy, June 1995, pp. 414–425.

[7] B. Nayfeh, L. Hammond, and K. Olukotun, "Evaluation of Design Alternatives for a Multiprocessor Microprocessor," in *Proc. of 23rd Annual Int. Symp. Computer Architecture*, Philadelphia, PA, June 1996, pp. 66–77.

[8] M. Franklin and G. Sohi, "ARB: a hardware mechanism for dynamic reordering of memory references," *IEEE Trans. Comput.*, Vol. 45, No. 5, May 1996, pp. 552–571.

[9] S. Gopal, T. N. Vijaykumar, J. E. Smith, and G. S. Sohi, "Speculative versioning cache," in *Proc. 4th Int. Symp. High-Performance Computer Architecture (HPCA-4)*, Las Vegas, NV, February 1998.

[10] N. P. Jouppi, "Improving direct-mapped cache performance by the addition of a small fully-associative cache and prefetch buffers," in *Proc. 17th Annu. Int. Symp. Computer Architecture*, Seattle, WA, June 1990, pp. 364–373.

[11] S. Muchnick, *Advanced Compiler Design and Implementation*. San Mateo, CA: Morgan Kaufmann, 1997.

[12] K. Olukotun, L. Hammond, and M. Willey, "Improving the performance of speculatively parallel applications on the hydra CMP," in *Proc. 1999 Int. Conf. Supercomputing*, Rhodes, Greece, June 1999, pp. 21–30.

[13] K. Olukotun, B. A. Nayfeh, L. Hammond, K. Wilson, and K. Chang, "The case for a single chip multiprocessor," in *Proc. 7th Int. Conf. Architectural Support for Programming Languages and Operating Systems (ASPLOS-VII)*, Cambridge, MA, Oct. 1996, pp. 2–11.

[14] M. Tremblay, "MAJC[TM]: an architecture for the new millennium," in *Hot Chips XI*, Stanford, CA, Aug. 1999, pp. 275–288. http://www.hotchips.org/archives/. Also reported in B. Case, "Sun makes MAJC with mirrors," *Microprocessor Report*, Oct. 25, 1999, pp. 18–21.

[15] L. Hammond, M. Willey, and K. Olukotun, "Data speculation support for a chip multiprocessor," in *Proc. 8th Int. Conf. Architectural Support for Programming Languages and Operating Systems (ASPLOS-VIII)*, San Jose, CA, Oct. 1998, pp. 58–69.

[16] J. G. Steffan et al., "A scalable approach to thread-level speculation," in *Proc. 27th Ann. Int. Symp. Computer Architecture (ISCA-27)*, Vancouver, BC, Canada, June 2000, pp. 1–12.

[17] T. Wilkinson, Kaffe Virtual Machine, http://kaffe.org, 1997–2002.

[18] http://www.byte.com

[19] http://www.specbench.org/jvm98/

[20] http://www.epcc.ed.ac.uk/javagrande/

CHAPTER 4

Improving Latency Using Manual Parallel Programming

Fully automatic systems such as TLS can allow us to utilize the multiple cores within a CMP to accelerate single applications to a certain extent, but realistically human programmers will always be able to do a better job at dividing up applications into separate tasks that can work efficiently on each of the cores within a CMP. However, historically speaking parallel programming has been so much more difficult than conventional uniprocessor programming that few have bothered to master its intricate difficulties. Communication between processors was generally slow in relation to the speed of individual processors, so it was critical for programmers to ensure that threads running on separate processors required only minimal communication between each other. Because reducing communication to manageable levels is often difficult, only a small minority of users bothered to invest the time and effort required to parallelize their programs in a way that could achieve speedup, and so these techniques were only taught in advanced, graduate-level computer science courses. Instead, in most cases programmers found that it was just easier to wait for the next generation of uniprocessors to appear and speed up their applications for "free" instead of investing the effort required to parallelize their programs. As a result, multiprocessors had a hard time competing against uniprocessors except in very large systems, where the target performance simply exceeded the performance obtainable by using even the fastest uniprocessors.

However, today parallel programming simply cannot be ignored. First, there is the "stick" of a slowing to the performance that programmers can expect to get for "free" from their uniprocessors. With the exhaustion of essentially all performance gains that can be achieved using technologies such as superscalar dispatch and pipelining, programmers must now actively switch to more parallel programming models in order to continue to increase the performance of their programs. Looking at the big picture, there are only three real "dimensions" to processor performance increases: clock frequency, superscalar instruction issue, and multiprocessing. The first two have been pushed to their logical limits, and so designers must now embrace multiprocessing, even if it means that programmers will be forced to change to a parallel programming model to achieve the highest possible performance.

Conveniently enough, CMPs offer a matching "carrot" effect that makes the transition from uniprocessor programming to parallel programming much easier than was possible in the past. Previously it was necessary to minimize communication between independent threads to an *extremely* low level, because each communication could require hundreds or even thousands of processor cycles. Also, proper management of communication within machines required extensive knowledge of the machine's communications architecture, a serious handicap for novice parallel programmers. Within a CMP with a shared on-chip cache memory, however, each communication event is a simple movement of cache lines from one processor to another and typically takes just a handful of processor cycles. With latencies like this, communication delays have a much smaller impact on overall system performance. Programmers must still divide up their work into parallel threads, but do not need to worry much about ensuring that these threads are highly independent, since communication is relatively cheap. Parallel threads can also be much smaller and still be effective—threads that are only hundreds or a few thousand cycles long can often be used to extract parallelism with these systems, instead of the millions of cycles-long threads typically necessary with conventional parallel machines. Researchers have shown that parallelization of applications can be made even easier with several schemes involving the addition of *transactional* hardware to a CMP [1–5]. These systems add buffering logic that lets threads *attempt* to execute in parallel, and then dynamically determines whether or not they are actually parallel at runtime. If no interthread dependences are detected at runtime, then the threads complete normally. However, if dependences exist, then buffers of some threads are cleared and those threads are restarted, dynamically serializing the threads in the process. Such hardware, which is really only practical on tightly coupled parallel machines such as CMPs, eliminates the need for programmers to even determine whether threads are parallel as they are parallelizing their programs—they need only choose *potentially* parallel threads. Overall, due to the short communication latencies and with enhancements like transactional memory, for programmers the shift from conventional processors to CMPs should be much less traumatic than the shift from conventional processors to multichip multiprocessors, finally moving within the range of what is feasible for "typical" programmers to handle.

While it applies equally well to CMPs, conventional parallel programming practices have been covered in numerous other texts, so the remainder of this chapter will focus on the most important new parallel programming technology that is really enabled by the tight connections between processors in a CMP: *transactional memory*.

4.1   USING TLS SUPPORT AS TRANSACTIONAL MEMORY

The TLS system described previously can also be regarded as a kind of transactional memory system for use with manual parallelization efforts, instead of as the basis for a fully automatic parallelization system. The key addition is that the system must be able to provide reports to programmers specifying where violations occur, since these indicate dependences between parallel threads. This provides a simple way to parallelize programs effectively, in a pseudo-sequential environment that prevents most common parallel programming errors and gives a programmer explicit feedback to direct his or her performance tuning efforts.

4.1.1   An Example: Parallelizing Heapsort Using TLS

In this section, a simple example application is used to illustrate many important points about how a programmer can use TLS to parallelize applications. Many of these techniques are also applicable for use with other transactional systems.

The example is C code that implements the main algorithm for a heap sort. In this algorithm, an array of pointers to data elements is used to sort the elements. Encoded in memory as a simple linear array (Fig. 4.1(a)), the node array is actually interpreted as a balanced binary tree by the algorithm (Fig. 4.1(b)). Tree sibling nodes are recorded consecutively in the array, while child nodes are stored at indices approximately twice that of their parents. For example, Node 2 is located directly after its sibling (Node 1) in the array, while the children of Node 2 (Nodes 5 and 6) are located adjacent to each other with indices approximately twice that of Node 2. This structure allows a complete binary tree to be recorded without requiring explicit pointers to connect parent and child nodes together, because the tree structure can

A) Tree structure in memory

Node	0	1	2	3	4	5	6	7	8	9	10	11	12	13	14
Address stored	A4	A2	A6	A1	A3	A0	A5	null	null	null	null	null	null	null	null

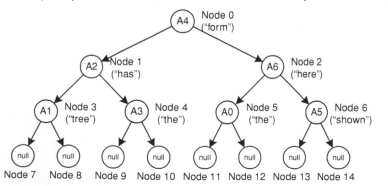

FIGURE 4.1: Organization of the heap array.

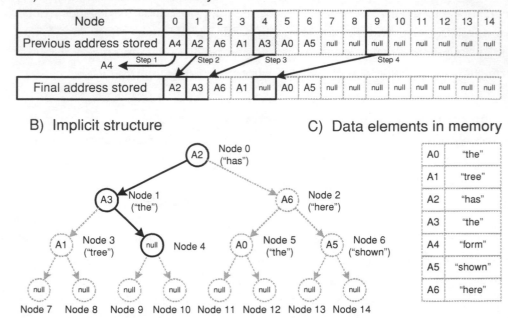

FIGURE 4.2: Top node removal and update of the heap.

always be determined arithmetically. In this example, each node of the tree consists of a single pointer to a variable-length data element located elsewhere in memory (Fig. 4.1(c)).

The heap is partially sorted. The element pointed to by any parent is always less than the element pointed to by each of the children, so the first pointer always points to the smallest element. Nodes are added to the bottom of the tree (highest indices) and bubble upward, switching places with parents that point to greater-valued elements. Final sorting is conducted by removing the top node (first pointer) and iteratively filling the vacancy by selecting and moving up the child pointer that points to the lesser element (Fig. 4.2). This example focuses only on this final sorting, which typically dominates the execution time of heap sort.

The code is provided in Fig. 4.3. It can be used to count the number of appearances of each (linguistic) word in a passage of text. It has been optimized for uniprocessor performance, so that parallelization with TLS can only derive speedups due to true parallelism and not due to more efficient code design. The code processes the preconstructed heap `node[]`, where each node (e.g., node[3]) is a pointer to a string (line 2). As each top node is removed and replaced from the remaining heap, a count is kept of the number of instances of each string dereferenced by the nodes (line 17). Each string and its count are written into a (previously allocated) result string (line 2) at the position pointed to by `inRes` (lines 9–16). To do this, the top node of the

```
1:  #define COLWID (30)
2:  char *result, *node[];

3:  void compileResults() {
4:    char *last, *inRes;
5:    long cmpPt, oldCmpPt, cnt;
6:    int sLen;

       //   INITIALIZATION
7:    inRes = result; last = node[0]; cnt = 0;

       //   OUTER LOOP - REMOVES ONE NODE EACH ITERATION
8:    while (node[0]) {
         //   IF NEW STRING, WRITE LAST STRING AND COUNT
         //   TO RESULT STRING AND RESET COUNT
9:      if (strcmp(node[cmpPt=0], last)) {
10:       strcpy(inRes, last);
11:       sLen = strlen(last);
12:       memset(inRes+sLen, ' ', COLWID-sLen);
13:       inRes += sprintf(inRes+=COLWID, "%5ld\n", cnt);
14:       cnt = 0;
15:       last = node[0];
16:     }

17:     cnt++;

         //INNER LOOP - UPDATE THE HEAP, REPLACE TOP NODE
18:     while (node[oldCmpPt=cmpPt] != NULL) {
19:       cmpPt = cmpPt*2 + 2;
20:       if (node[cmpPt-1] && !(node[cmpPt] &&
            strcmp(node[cmpPt-1], node[cmpPt]) >= 0))
21:         --cmpPt;
22:       node[oldCmpPt] = node[cmpPt];
23:     }
24:   }

       //   WRITE FINAL STRING AND COUNT TO RESULT STRING
25:   strcpy(inRes, last);
26:   sLen = strlen(last);
27:   memset(inRes+sLen, ' ', COLWID-sLen);
28:   sprintf(inRes+=COLWID, "%5ld\n", cnt);
29: }
```

FIGURE 4.3: Code for top node removal and heap update.

heap (node[0], which points to the alphabetically first string) is removed and compared to the string pointed to by the previous top node removed (lines 8 and 9). If they point to dissimilar strings, then all nodes pointing to the previous string have been removed and counted, so the string and its count are written to the result string and the count is reset (lines 9–16). In all cases, the count for the current string is incremented (line 17) and the heap is updated/sorted in the manner described above (lines 18–23). The heap is structured so that below the last valid child on any tree descent, the left and right child are always two NULL pointer nodes (line 18). This whole counting and sorting process is conducted until the heap is empty (line 8). Then the results for the last string are written to the result string (lines 25–28).

4.1.1.1 Manually Parallelizing with TLS

When manually parallelizing with TLS, the programmer first looks for parts of the application with good parallel qualities, as was discussed in the last chapter. These parts should dominate the execution time of the application with the time concentrated in one or more loops, preferably with a number of iterations equal to or greater than the number of processors in the TLS CMP. These loops should contain fairly independent tasks (few intertask data dependences), with each task requiring from 200 to 10,000 cycles to complete, and all tasks being approximately equal in length for good load balancing. For the example program, the two loop levels where a programmer can parallelize this code are either the inner loop or the outer loop, i.e., within a single event of removing node[0] and updating the heap (lines 8–24), or across multiple such events. The first is not good due to the small parallel task sizes involved, which are too small to amortize the overhead required to create new threads. The second level is much better suited to the per-iteration overheads of the TLS system. But, parallelizing across multiple node removals and heap updates requires each thread to synchronize the reading of any node (lines 8, 9, 15, 18, 20, 22) with the possible updates of that node by the previous threads (line 22). The top node will always require synchronization, while nodes at lower levels will conflict across threads with a decreasing likelihood at each level.

One can perform an initial parallelization using TLS simply by choosing and specifying the correct loop to parallelize using something like a special keyword, compiler #pragma, or the like. The compiler must then ensure that the initial load from and the final store to shared variables or to dereferenced pointers within each thread occurs in a way that is visible to the TLS hardware, usually by forcing the data moves through memory at these borders. This ensures that the TLS system can detect all interthread data dependence violations, when necessary. Meanwhile, all variables without loop-carried dependences are made private to each thread to prevent false sharing and violations.

As a test, this code was executed upon a heap comprising the approximately 7800 words in the U.S. Constitution and its amendments, while running on the Hydra CMP with speculation support described in the previous chapter. Without further modification, the TLS CMP provided a speedup of 2.6 over a single-processor system with the same, unscaled, realistic memory system. Most of the performance loss from the "perfect" speedup of 4 was due to true dependences among the threads; the requirement that shared variables not be register-allocated across thread boundaries caused only a 2% slowdown. In the remainder of this section, this level of performance is described as the "base" TLS parallelization.

4.1.1.2 Ease of TLS Parallelization

The base case illustrates the simplicity of transactional programming with TLS hardware, in contrast to the complexity and overheads of conventional parallelization, which would require extensive locking even for this simple example. Like TLS, conventional parallelization requires that loop-carried dependences be identified. However, once this has been done, the difficult part of conventional parallelization begins. Accesses to any dereferenced pointer or variable with loop-carried dependences could cause data races between processors executing different iterations in parallel. While synchronization must be considered for each access, to avoid poor performance only accesses that could *actually* cause data races should be synchronized with each other. However, determining which accesses conflict requires either a good understanding of the algorithm and its use of pointers or a detailed understanding of the memory behavior of the algorithm. Pointer aliasing and control flow dependences can make these difficult. Finally, a locking scheme must be devised and implemented. This typically requires changes in the data structures or algorithms and must be carefully considered to provide good performance. None of this is necessary when parallelizing with TLS.

In this example, one set of accesses that must be explicitly synchronized when using conventional parallelization are the read accesses of the nodes (lines 8, 9, 15, 18, 20, 22) with the possible updates of those nodes by earlier iterations (line 22). To do this a new array of locks could be added, one for each node in the heap. However, this would introduce large overheads. Extra storage would be required to store the locks. Each time a comparison of child nodes and an update of the parent node were to occur, an additional locking and unlocking of the parent and testing of locks for each of the child nodes would need to be done. Furthermore, doing this correctly would require careful analysis. The ordering of these operations would be critical. For example, unlocking the parent before locking the child to be transferred to the parent node would allow for race conditions between processors. Worse yet, these races would be challenging to correct because they would be difficult to detect, to repeat and to understand.

One could attempt a different synchronization scheme to lower the overheads. For example, each processor could specify the level of the heap that it is currently modifying, and processors executing later iterations could be prevented from accessing nodes at or below this level of the heap. While this would reduce the storage requirements for the locks to just one per processor, it would introduce unnecessary serialization between accesses to nodes in different branches of the heap. Another alternative would be to have each processor specify only the node which is being updated, so processors executing later iterations would stall only on accesses to this node. However, locking overheads would still exist in either case, and care would still need to be taken to prevent data races. Alternatively, the choice could be made to completely replace the uniprocessor heap sort with a new algorithm designed for parallelism from the start. However, this would likely be more complex than any solution discussed so far, and the support for parallelism will still introduce overheads into any algorithm that has interthread dependences. As this example shows, parallelization without TLS can be much more complex and error prone than parallelization with TLS. Because the complexity of redesign versus incremental modification becomes greater for larger, more complex programs, its simplicity is even more of a benefit for real-world applications.

In addition, performance of an algorithm like this parallelized using TLS can often be better than a conventionally parallelized one, primarily because it is often possible to *optimistically* speculate beyond potential dependences, eliminating all synchronization stall time when the potential violations do not actually occur. It can be much more efficient than the pessimistic static synchronization used in conventional parallelization, which synchronizes on all possible dependences, no matter how unlikely. In fact, TLS can often improve the performance of an application that has *already* been manually parallelized by allowing some optimistic parallelization [6]. Less apparent is that a single-threaded application only incrementally modified using manual TLS parallelization can sometimes provide better performance than an application that has been completely redesigned for optimal parallel performance using only conventional manual parallelization. This is because code optimized for non-TLS parallel performance introduces overhead over uniprocessor code to support low-contention parallel structures, algorithms and synchronization. The advantage that results from this redesign for conventional parallelism can be less than the combined advantages of using TLS and starting with more efficient, optimal uniprocessor code. Given the difficulty of redesigning legacy code and of parallel programming, this can make TLS with manual enhancements a better alternative than using conventional manual parallelization.

4.1.1.3 Optimizing TLS Parallel Performance

To optimize code parallelized using TLS, a programmer conducts the base TLS parallelization, as described above, and then executes the resultant code against a representative data set. TLS hardware should include a performance tuning mode that provides reporting of dependence violations, including data on which processors were involved in the violation, the address of the violating data element, which load and store pairs triggered the violation, and how much speculative work was discarded. This data is then sorted by each load–store violation pair. This data is the same as that used by Jrpm during automated profiling. By totaling the cycles discarded for each pair and sorting the pairs by these totals, the causes of the largest losses can be quickly assessed. Using this ranking, a programmer can better understand the dynamic behavior of the parallel program and more easily reduce violation losses.

Compared to non-TLS parallel programming, manual parallelization with TLS allows the programmer to more quickly transform a portion of code. The key to this is that TLS provides the ability to easily test the dynamic behavior of speculatively parallel code (while it correctly executes in spite of dependence violations) and get specific information about the violations most affecting performance. The programmer can then focus only on those violations that most hamper performance, rather than being required to synchronize each potentially violating dependence to avoid introducing data races into the program.

Typically, parallel performance is most severely impacted by a small number of interthread data dependences. Moving the writes as early as possible within the less speculative thread and the reads as late as possible within the more speculative thread usually reduces the chance of experiencing a data dependence violation. For loop-based TLS, this corresponds to moving performance-limiting writes toward the top of the loop and delaying performance-limiting reads toward the end of the loop; in the limit, the first load of a dependent variable occurs just prior to the last store, forming a tiny critical region. Furthermore, moving this critical region as close as possible to the top of the loop minimizes the execution discarded when violations do occur. Finally, constructing the loop body to ensure that the critical region always occurs approximately the same number of cycles into the execution of the loop and requires a fairly constant time to complete allows the speculative threads to follow each other with a fixed interthread delay without experiencing violations. In contrast, critical sections that occur sometimes early and sometimes late increase violations due to late stores in less speculative threads violating early reads in more speculative ones.

4.1.1.4 Manual Code Motion

In the example, more than three violations per committed thread occur while executing the base parallelization. The store of `last` in line 15 often violates the speculative read of `last` in line 9. The same occurs with `cnt` (the store in line 17 violates the load in line 13), `inRes`, and several other variables. To reduce these violations, one can minimize the length of the critical regions from first load to last store. For example, the store of last in line 15 can be moved right after the load in line 9. Because each thread optimally executes with a lag of one-quarter iteration from the previous thread on a four-processor CMP, this makes it unlikely that any other thread will be concurrently executing the same critical region. To hoist the store of `last`, the previous value must first be saved in a temporary variable for lines 10 and 11. Unlike most other modifications discussed below, research shows that this transformation can be automated in many cases [7]. It is also a good idea to move these critical regions as early in each thread as possible. For example, line 17 (the increment of `cnt`) can be moved above the conditional block (lines 9–16).

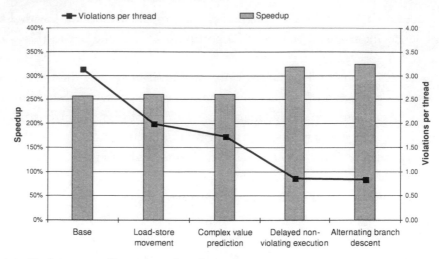

FIGURE 4.4: Performance of incremental optimizations.

When these transformations have been completed for all variables that can benefit, surprisingly the performance remains virtually unchanged. Most of the lines which were causing violations before are no longer significant sources of losses, but now previously unimportant load–store violation pairs have "popped up" and now dominate performance by causing much larger losses than before. Threads now progress farther per violation, but nonetheless violate anyway before they can successfully commit their results. This results in a lower violation count, but more discarded execution time per violation. This is shown in Fig. 4.4, which shows speedup results and the number of violations per committed thread for each version of the example application.

Unfortunately, the performance at this point (a speedup of 2.6) represents an optimistic upper bound on the current capability of simple code motion. However, it is possible to get still more speedup (a final speedup of 3.4) with a minimum of code transformation. This is because a programmer can do things such as more complex value prediction, involving minor changes to data structures or algorithms, in addition to simple code motion. These additional changes require an increasingly detailed understanding of the application.

4.1.1.5 Complex Value Prediction

In the current example, one of the main variables suffering violations is `inRes`. Complex value prediction can reduce these violations. Note that the result string is constructed out of fixed width columns. The first column is `COLWID` characters wide and contains the word (lines 10–12). The next column is five characters wide and contains the final count of the number of instances of the string, followed by a carriage return (line 13). From the code a programmer can determine that the final value of inRes will always be COLWID+5+1 characters greater after line 13 than it was in line 10. Using this prediction of the final value of inRes, the programmer is able to hoist the final update of inRes above the many function calls in lines 10–13, once again reducing both the chance of a violation occurring and the execution time discarded if a violation does occur.

Likewise, if the count had been printed to a variable-length field, the programmer could have chosen to change the format to a fixed length to allow for complex value prediction. This could occur if the output format was not critical and could tolerate a change. If so, this would also show how a small change in the algorithm and data structures can allow further optimization on a program exhibiting contention due to its having been designed without parallel execution in mind.

4.1.1.6 Algorithm Adjustments

By this point almost all loads and stores to the same variable are placed close to each other and close to the top of each iteration, and yet the performance has not improved significantly. Upon closer examination, it is apparent that many of the violations would never occur if each thread did not execute lines 9–16 and if it maintained a spacing of one-quarter iteration from the threads immediately previous to and following it. The problem is that when lines 9–16 are executed, a large number of cycles are consumed to store a word and its count to the result string. Only after completing this, the thread updates the top of the heap (line 22). This violates all more speculative processors, due to the load in line 8, and causes them to discard all their execution during the time the result string was being updated. While conducting an early update on the top node of the heap could yield some benefit, nodes further down would still likely cause violations.

The optimization to alleviate this problem is to move as much of the execution in lines 9–16 to the position following line 23. By minimizing the work conducted before lines 17–23, one can reduce or eliminate many of the violations. In particular, only the updates of data locations with loop-carried dependences should occur before line 17, i.e., updates to inRes, last, and cnt. The strcpy, strlen, memset, and sprintf functions can be conducted later, after lines 17–23, without causing violations. This is similar to moving load–store pairs closer to the start of each iteration, but instead the programmer is making algorithm changes to move nonviolating work closer to the end of each iteration. Specifically, these four time-consuming functions are moved from before to after the heap update, which repeatedly dereferences dynamically determined pointers. As Fig. 4.4 shows, this optimization greatly improves performance, raising the speedup from 2.6 to 3.2 and also halving the number of violations.

After this optimization, the dominant remaining violations are the loads in line 20 with the store in line 22. This is often due to the fact that when the two child nodes point to equal strings (a common occurrence at the top of the heap), the second (right, higher index) node is always selected. This leads to frequent contention for all nodes near the top of the heap and resultant violations, as each thread descends down the same path through the heap. One can easily change the algorithm so that each speculative thread chooses the opposite direction from the thread immediately before it. Consecutive threads will alternate between always selecting the left and always selecting the right node in cases of equality, thereby descending down the opposite path from the immediately previous thread. This alters the behavior of the program somewhat, and therefore requires additional work to ensure that the program will still produce exactly the same final result string—which it does. This final optimization results in slightly improved performance and less frequent violations. Note that including overhead code from all the transformations so far would yield a 4% slowdown if the code were executed on a uniprocessor. Hence, the most one can hope for would be a speedup of 3.85 versus the original sequential program.

Further attempts at optimization were unsuccessful. Violations do remain, but they occur infrequently enough that their occasional losses are less than the overhead of executing code devoted to eliminating them. For example, attempts at synchronizing on the most frequent violations, using locks like those described in the last chapter, simply generated excessive waiting times.

4.1.2   Parallelizing SPEC2000 with TLS

This approach was evaluated using seven benchmarks from the SPEC2000 suite: four floating point applications that are coded in C, since they are more difficult to parallelize than the Fortran benchmarks, and three of the integer benchmarks, selected based upon their source code size and indications from profiling that they would be amenable to manual parallelization with TLS. For example, a high concentration of execution time within just a small number of functions was considered a good sign. Information on the selected benchmarks is given in Table 4.1.

These benchmarks were run using the reference input data sets. Due to the long execution times of these data sets, complete execution was not possible for any of the benchmarks. Since previous research on SPEC benchmarks [8] has demonstrated both the difficulty and the importance of carefully choosing the portion of execution to simulate for applications that exhibit large-time-scale cyclic behavior, one or more whole application cycles were simulated in all cases. The total of all simulation samples was at least 100 million instructions from each original (nonparallelized) application. One should note that all speedup and coverage results presented below are based upon an extrapolation of these samples of whole

TABLE 4.1: Benchmark characteristics

BENCHMARK		APPLICATION CATEGORY	LINES OF CODE
CFP 2000	177.mesa	3-D graphics library	61,343
	179.art	Image recognition/neural networks	1,270
	183.equake	Seismic wave propagation simulation	1,513
	188.ammp	Computational chemistry	14,657
CINT 2000	175.vpr	FPGA circuit placement and routing	17,729
	181.mcf	Combinatorial optimization	2,412
	300.twolf	Place and route simulator	20,459

application cycles back to the entire application. The extrapolation was conducted by first profiling the full sequential application using real hardware and the same compiler. Full application speedup was then calculated assuming that the simulated speedup results were representative of entire phases of similar code, while assuming no speedup would occur during other program phases.

4.1.2.1 Results of Parallelization

Each application was initially parallelized using base parallelization of loops and simple code motion, and then modified further in a variety of ways to show additional benefits. Table 4.2 lists the additional transformations that were then used. The first three are the ones that are simple enough so that they could conceivably be automated; the second three definitely require manual intervention. The data demonstrate that the simple transformations are beneficial for both floating point and integer applications. However, the complex ones are beneficial mainly for the integer applications. This was because the execution times of the floating point applications were all dominated by easily parallelizable loops, except for *ammp*. Therefore, the complex transformations added little or no benefit. In contrast, all the integer applications benefited from the complex code transformations.

Table 4.3 details the speedups achieved for each application as the transformations were sequentially added. Ideally, the incremental speedup due to each transformation could be listed. However, the transformations interact with each other. For example, on *vpr* (place) explicit synchronization yielded no speedup after base parallelization with additional value prediction. However, applying it together with the parallel reduction transformation provided a sizeable

TABLE 4.2: Code transformations applied

TRANSFORMATION	SPEC CFP2000				SPEC CINT2000		
	177 MESA	179 ART	183 EQUAKE	188 AMMP	175 VPR	181 MCF	300 TWOLF
Loop chunking/slicing	X	X	X		X		
Parallel reductions		X			X	X	X
Explicit synchronization					X		X
Speculative pipelining				X	X	X	X
Adapt algorithms or data structures					X	X	X
Complex value prediction					X	X	X

advantage. Due to the interactions and the many permutations of transformations, speedups are only listed for a single sequence of transformations. More information about the exact optimizations can be found in [9]. Note that because *vpr* is a place and route application and the two portions of the application are very different, results for them are listed separately.

4.2   TRANSACTIONAL COHERENCE AND CONSISTENCY (TCC): MORE GENERALIZED TRANSACTIONAL MEMORY

While TLS systems offer a form of transactional memory that can be effectively used by parallel programmers to manually enhance performance over "automatic" levels, true TLS systems are hard to build with more than a few processors, because they require high snoop bandwidth among the processors on the CMP. Practically speaking, it would be difficult to scale beyond about 8–16 processors, and the bandwidths are definitely too high to allow TLS to occur between chips in a multi-CMP system. Hence, an obvious question is whether or not it is possible to develop a more scalable system that retains most of the advantages of TLS from a parallel programmer's point of view.

To solve this, one must identify the key advantages of parallelization with TLS are over conventional parallelization. The first is the "pseudo-sequential" programming model provided by the fact that each thread must complete as an atomic operation, so the hardware guarantees

TABLE 4.3: Summary of optimization results.

Application	Speculative rgns.	Location of top level of speculative region(s). Line numbers are for SPEC CPU2000, version 1.00.	Percent execution time coverage	Last transformation applied	Real memory Incremental speedup	Real memory Cumulative speedup	Perfect memory Incremental speedup	Perfect memory Cumulative speedup	Perfect memory, no spec overhead Incremental speedup	Perfect memory, no spec overhead Cumulative speedup
177. mesa	1	vbrender.c, lines 897–901	84%	Basic	175%	175%	179%	179%	179%	179%
179. art	7	scanner.c, lines 405–477 (7 loops) and 545–617 (7 loops)	95%	Basic	60%	60%	0%	0%	0%	0%
				Parallel reductions	39%	122%	43%	43%	112%	112%
				Loop chunking/slicing	14%	154%	167%	282%	86%	294%
183. equake	6	quake.c, lines 449–478 (5 loops) and 1195–1220	100%	Basic	135%	135%	185%	185%	195%	195%
				Loop chunking/slicing	4%	145%	4%	196%	2%	200%
188. ammp	1	rectmm.c, lines 562–1123	86%	Basic	61%	61%	59%	59%	62%	62%
				Speculative pipelining, loop chunking/slicing	24%	99%	6%	69%	9%	76%
175. vpr (place)	1	place.c, lines 506–513	100%	Basic	7%	7%	16%	16%	17%	17%
				Complex value prediction	45%	55%	44%	67%	44%	68%
				Parallel reductions, explicit synchronization	36%	111%	37%	128%	36%	129%
175. vpr (route)	1	route.c, lines 518–541	97%	Speculative pipelining	17%	17%	16%	16%	27%	27%
				Algorithm/data structure modifications	43%	67%	38%	60%	35%	72%
				Complex value prediction	13%	88%	33%	113%	33%	128%
181. mcf	6	implicit.c, lines 246–272	44%	Loop chunking/slicing, algorithm/data structure changes	70%	70%	126%	126%	151%	151%
		mcfutil.c, lines 75–76	5%	Loop chunking/slicing, complex value prediction	24%	24%	>300%	>300%	>300%	>300%
		mcfutil.c, lines 80–109	19%	Parallel reductions, speculative slices, speculative pipelining, complex value prediction	55%	55%	10%	10%	16%	16%
		pbeammp.c, lines 96–121	7%	Speculative pipelining, algorithm/data structure changes	84%	84%	95%	95%	119%	119%
		pbeammp.c, lines 161–174	4%	Basic	64%	64%	146%	146%	197%	197%
		pbeammp.c, lines 181–195	20%	Loop chunking/slicing	89%	89%	150%	150%	211%	211%
300. twolf	1	uloop.c, lines 154–361	100%	Speculative pipelining	18%	18%	21%	21%	23%	23%
				Parallel reductions, explicit synchronization, algorithm/data structure changes	21%	43%	26%	53%	29%	59%
				Complex value prediction	12%	60%	9%	67%	8%	72%

CFP2000

CINT2000

that all of its loads and stores appear to execute after all "earlier" threads have completed and before all "later" threads have completed, even though those threads are actually executing at the same time. This makes the parallel programming experience a much more gradual step up from sequential programming by eliminating the need for programmers to master the complexity of locks, messages, and the like. Secondarily, the automatic violation detection when dependence violations occur ensures that programs will always execute correctly, even if the programmer chooses threads poorly. The program may not speed up, but at least they will still get the correct answer. Traditional parallel programming has effectively required that programmers break their original program and rebuild it to a new specification, a strategy that was often error prone. Automatic violation detection also provides a way to let the machine find all dependences for you and classify them according to how serious they are, directing programmers right to the "trouble spots" in their program instead of forcing them to guess as to which parts of their program really need optimization. Other aspects of TLS, such as immediate forwarding of writes to subsequent loads, are helpful but not critical. Hence, one way to make the system more scalable is to eliminate the forwarding and continuous broadcasts of all writes. Also, to prevent load imbalance from becoming a major bottleneck as the number of processors scales, one can keep the hardware-sequenced commit mechanism, but allow software to loosen its strict one-commit-order rule when possible. With these adjustments to TLS, one can build a new system: transactional coherence and consistency (TCC). For programmers, this system has most of the advantages of TLS, but it is more scalable and flexible due to its simpler hardware requirements and more flexible definition of parallel threads.

4.2.1   TCC HARDWARE

Processors operating in a TCC-based multiprocessor *continually* execute speculative transactions, using a cycle illustrated in Fig. 4.5(a) on multiprocessor hardware with additions similar to those depicted in Fig. 4.5(b). A transaction in a TCC system is a sequence of instructions marked by software that is guaranteed to execute and complete only as an atomic unit, and acts much like a TLS thread. Each transaction produces a block of writes which are buffered locally while the transaction executes and are then committed to shared memory *only* as an atomic unit, after the transaction completes. Unlike TLS systems, writes are not broadcast until after each transaction has completed execution. Once the transaction is complete, hardware must arbitrate system-wide for the permission to commit its writes. After this permission is granted, the processor can take advantage of high-bandwidth system interconnect to broadcast all writes for the entire transaction out as one packet to the rest of the system. This broadcast may make scaling TCC to immense numbers of processors a challenge, but it is still much simpler to broadcast blocks of writes from each processor than it is to continually broadcast individual writes throughout the system, and more amenable to optimizations that can reduce the amount

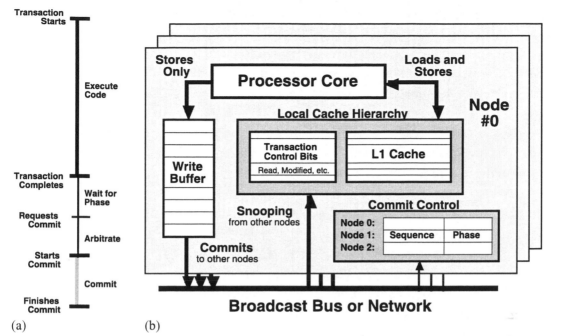

FIGURE 4.5: (a) A transaction cycle (time flows downward), and (b) a block diagram of sample TCC-enabled hardware.

of system-wide traffic as each commit occurs. Meanwhile, the local caches in other processors snoop on these store packets to maintain coherence in the system, in a manner analogous to snooping on TLS writes. Snooping also allows them to detect when dependence violations occur.

TCC's hardware requirements also compare favorably with conventional parallel systems, because the hardware required to support the illusion of a shared memory in conventional parallel processors can be very complex. In order to provide a coherent view of memory, the hardware must track where the latest version of any particular memory address can be found, recover the latest version of a cache line from *anywhere* on the system when a load from it occurs, and efficiently support the communication of large numbers of small, cache-line-sized packets of data between processors. All this must be done with minimal latency, too, since individual load and store instructions are dependent upon each communication event. In contrast, TCC's ability to combine all writes from the entire transaction together imparts latency tolerance, because fewer interprocessor messages and arbitrations are required, and because flushing out the writes is a one-way operation. Further complicating matters is the problem of *sequencing* the various communication events constantly passing throughout a conventional parallel system at the fine-grained level of individual load and store instructions. Hardware rules known as memory consistency models must be devised and correctly implemented to allow software synchronization routines to work reliably. Over the years, these models have progressed from the easy-to-understand but sometimes performance-limiting sequential consistency to more modern schemes such as relaxed consistency. With TCC, the commit operation can also be leveraged to provide sequencing between memory operations—organized by thread—and hence a greatly simplified consistency protocol.

The continual cycle of speculative buffering, broadcast, and (potential) violations used by TCC allows one to replace both conventional coherence and consistence protocols with a much simpler model:

Consistence. By only controlling ordering between transaction commits, instead of individual loads and stores, TCC drastically reduces the number of latency-sensitive arbitration and synchronization events required by low-level protocols in a typical multiprocessor system. It can also simplify the design of parallelizing compilers, which can orchestrate communication more precisely. Imposing an order on the transaction commits and backing up uncommitted transactions if they have speculatively read data modified by other transactions effectively lets the TCC system provide an illusion of uniprocessor execution to the sequence of memory references generated by software. As far as the global memory and software is concerned, all memory references from a transaction that commits earlier happened "before" all of the memory references of a transaction that commits afterwards, even if their actual execution was interleaved in time, because all writes from a transaction become visible to other processors only at commit time, all at once.

Coherence. Stores are buffered and kept within the processor node for the duration of the transaction in order to maintain the atomicity of the transaction. No conventional MESI-style protocols are used to maintain lines in "shared" or "exclusive" states at any point in the system, so it is legal for many processor nodes to hold the same line simultaneously in either an unmodified or speculatively modified form. At the end of each transaction, its commit broadcast notifies all other processors about what state it has changed. During this process, the other processors perform conventional invalidation (if the commit packet contains only addresses) or update (if it contains addresses and data) to keep their cache state coherent. Simultaneously, they must determine if they may have read from any of the committed addresses. If so, they are forced to restart and re-execute their current transactions with the updated data. This protects against true data dependences. At the same time, data antidependences are handled simply by the fact that later processors will eventually get their own turn to flush out data to memory. Until that point, their "later" results are not seen by transactions that commit earlier (avoiding write-after-read antidependences) and they are able to freely overwrite previously modified data in a clearly sequenced manner (handling write-after-write antidependences in a legal way). Effectively, the simple, sequentialized consistence model allows for a simpler coherence model, too.

TCC will work in a wide variety of multiprocessor hardware environments, including a variety of CMP configurations and small-scale multichip multiprocessors. Within these systems, individual processor cores and their local cache hierarchies must have some features to provide speculative buffering of memory references and commit arbitration control. The most significant addition is a mechanism for collecting all modified cache lines from each transaction together into a commit packet. This can be implemented either as a write buffer completely separate from the caches or as an address buffer that maintains a list of the line tags that contain data to be committed. This buffer needs to be able to hold 4–16 KB worth of cache lines. If it fills up during execution of a long transaction, the processor must declare an "overflow" and stall until it obtains commit permission, when it may continue while writing its results directly to shared memory while holding the permission. This holding of the commit permission can cause serious serialization if it occurs frequently, but is acceptable on occasion.

In the caches, all of the included lines must maintain read and modified bits that are analogous to the TLS ones, described in the last chapter. Other bits may also be added to improve performance, but are not essential for correct execution. As with TLS, cache lines with set read bits may not be flushed from the local cache hierarchy in mid-transaction, or the processor must declare an "overflow" condition and stall until it acquires permission to commit its results. Set modified bits will cause similar overflow conditions if the write buffer only holds addresses.

4.2.2   TCC Software

TCC parallelization requires only a few new programming constructs. Using them is simpler than parallelization with conventional threaded models because it reduces the number of code transformations needed for typical parallelization efforts. It is also more flexible than the TLS model, allowing programmers to make informed tradeoffs between programmer effort and performance. The process is much like parallelizing for TLS, but with one major addition: the programmers can now specify any desired sequencing order among their transactions. Like TLS, the default ordering for transactions is to have them commit results in the same order as the original sequential program, since this guarantees that the program will execute correctly. However, if a programmer is able to verify that this commit order constraint is unnecessary, then it can be relaxed completely or partially in order to provide better performance. The interface also provides ways to specify the ordering constraints of the application in useful ways.

The remainder of this section describes coding techniques for two different mechanisms analogous to those used to extract threads automatically using TLS: a loop-based mechanism and a thread-forking scheme. Both of these techniques mark transactions in code and, additionally, indicate how the code should be broken into threads. As a result, they are a good way to parallelize originally sequential programs while using transactions. The example interface, including a few helper functions, is described in C, but it can be readily adapted to any programming language. In addition, a simpler extension is available for marking transactional behavior in programs that have already been broken up into threads using conventional techniques.

4.2.2.1 Loop-Based Parallelization

The parallelization of loops will be introduced in the context of a simple sequential code segment that calculates a histogram of 1000 integer percentages using an array of corresponding buckets:

```
int* data = load_data(); /* input */
int i, buckets[101];
for (i = 0; i < 1000; i++) {
    buckets[data[i]]++;
}
print_buckets(buckets); /* output */
```

The compiler interprets this program as one large transaction, exposing no parallelism to the TCC hardware. One can parallelize the for loop, however, with a modified keyword such as t_for:

```
...
t_for (i = 0; i < 1000; i++) {
...
```

With this small change, the program is a parallel loop that is guaranteed to execute the original sequential loop correctly, just like with TLS. Similar keywords, such as t_while, may also be useful. Each iteration of the loop body will now become a separate transaction that can execute in parallel, but must commit in the original sequential order, in a pattern like that in Fig. 4.6(a). When two parallel iterations try to update the same histogram bucket simultaneously, the TCC hardware will cause the "later" iteration to violate when it lets the "earlier" one commit, forcing the "later" one to re-execute using updated data and preserving the original sequential semantics.

In contrast, a conventionally parallelized system would require an array of locks to protect the histogram bins, resulting in much more extensive changes:

```
int* data = load_data();
int i, buckets[101];
```

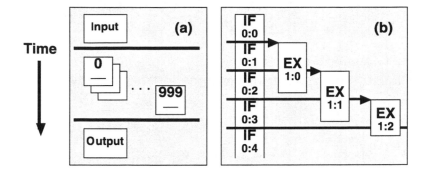

FIGURE 4.6: (a) For TCC loops, and (b) a simple fork example. The numbers in b's transactions are sequence:phase.

```
/* Define & initialize locks */
LOCK_TYPE bucketLock[101];
for (i = 0; i < 101; i++) {
  LOCK_INIT(bucketLock[i]);
}
for (i = 0; i < 1000; i++) {
  LOCK(bucketLock[data[i]]);
  buckets[data[i]]++;
  UNLOCK(bucketLock[data[i]]);
}
print_buckets(buckets);
```

Unlike the TCC version, if any of this locking code is omitted or buggy, then the program may fail—and not necessarily in the same place every time—significantly complicating debugging. Debugging is especially hard if the errors only happen for infrequently occurring memory access patterns. The situation is potentially even trickier if multiple locks need to be held simultaneously within a critical region, because one must be careful to avoid locking sequences that may deadlock.

Although sequential ordering is generally useful because it guarantees correct execution, in some cases—such as this histogram example—it is not actually required for correctness. In this case, the only dependences among the loop transactions are through the histogram bin updates, which can be performed in any order. When programmers can verify that all dependences are not order-critical, or if there are simply no loop-carried dependences, then they can use modified keywords such as t_for_unordered and t_while_unordered to allow the loop's transactions to commit in any order. Allowing unordered commits is most useful in more complex programs where the transaction lengths may vary dynamically, because it eliminates much of the time that processors spend stalled waiting for commit permission between unbalanced transactions.

4.2.2.2 Fork-Based Parallelization

While the simple parallel loop API will work for many programs, some less structured programs may need to generate transactions in a more flexible manner. For these situations one should use something more like conventional thread creation APIs, such as the t_fork call, below:

```
void t_fork(void (*child_function_ptr)(void*),
        void *input_data,
        int child_sequence_num,
        int parent_phase_increment,
        int child_phase_increment);
/* Which forks off a child function like: */
void child_function(void *input_data);
```

This call forces the "parent" transaction to commit, and then creates two completely new—and parallel—transactions in its place. One (the "new parent") continues execution of the code immediately following **t_fork**, while the other (the "child") starts executing the function at child_function_ptr with input_data. Other input parameters control ordering of forked transactions in relation to other transactions, and are discussed in more detail below. One can demonstrate this function with a parallelized form of a simple two stage processor pipeline. This is simulated using the functions i_fetch for instruction fetch, increment_PC to select the next instruction, and execute to execute instructions. The "child" transaction then executes each instruction while the "new parent" transaction fetches another:

```
/* Define an ID number for the EX sequence */
#define EX_SEQ 1

/* Initial setup */
int PC = INITIAL_PC;
int opcode = i_fetch(PC);

/* Main loop */
  while (opcode != END_CODE)
{
  t_fork(execute, &opcode, EX_SEQ, 1, 1);
  increment_PC(opcode, &PC);
  opcode = i_fetch(PC);
}
```

This example creates a sequence of overlapping transactions like those in Fig. 4.6(b). t_fork gives enough flexibility to divide a program into transactions in virtually any way. It can even be used to build the t_for and t_while constructs, if necessary.

4.2.2.3 Explicit Transaction Commit Ordering

The simple "ordered" and "unordered" ordering modes may not always be sufficient. For example, a programmer may desire partial ordering, executing unordered most of the time, but occasionally imposing some ordering. This is rare with transactional loops, but quite common with forked transactions.

Transaction ordering can be controlled by assigning two parameters to each transaction: the sequence and phase of transactions. These two numbers control the ordering of transaction commits. Transactions with the same sequence number may need to commit in a programmer-defined order, while transactions in different sequences are always independent. The t_fork call can be used to produce a "child" transaction with a new sequence number using the child_sequence_num parameter. Within each sequence, the phase number indicates the relative "age" of each transaction. TCC hardware will only commit transactions in the oldest active phase (lowest value) from within each active sequence. Using this notation, an ordered loop is just a sequence of transactions with the phase number incremented by one every time, while an unordered loop uses transactions all with the same phase number.

More arbitrary phase ordering of transactions can also be imposed with:

```
void t_commit(int phase_increment);
void t_wait_for_sequence(int phase_increment, int
    wait_for_sequence_num);
```

The t_commit routine implicitly commits the current transaction, and then immediately starts another on the same processor with a phase incremented by the phase_increment parameter. The most common phase_increment parameter used is 0, which just breaks a large transaction into two. However, it can also be used with a phase_increment of 1 or more in order to force an explicit transaction commit ordering. One use for this is to emulate a conventional barrier among all transactions within a sequence using transactional semantics. The similar t_wait_for_sequence call performs a t_commit and waits for all transactions within another sequence to complete. This call is usually used to allow a "parent" transaction sequence to wait for a "child" sequence to complete, like a conventional thread join operation.

4.2.2.4 Transactions in Prethreaded Code

Most of the complexity of the previous APIs was required not by their transactional nature, but by the mechanisms through which they automatically started off threads for each transaction, allowing all transactions to run in parallel even if the underlying software was originally written in a sequential manner. This is a great boon when one is attempting to parallelize sequential code using transactions, but is not necessary if the program has already been broken up into threads using conventional threading APIs.

For programs that are already parallel, it is possible to break up the existing threads into transactions by simply sprinkling `t_commit` calls throughout the code in such a way so that they never split code segments that *must* be executed atomically—typically, the critical regions surrounded by "lock" and "unlock" calls in most conventionally parallelized code—and also so that the `t_commits` are triggered on a regular basis, to avoid transactional hardware buffer overflows. For C-language programming, especially, this is a quick way to make parallel programs into transactional ones. For programmers, however, a slightly different API, based on Java `synchronized` blocks, makes the breakdown of the program into transactions more clear:

atomic {
        . . . your atomic block here . . .
}

By adding the new `atomic` keyword to your language, regions in the program that *must* be executed atomically can be marked in a very clear, block-oriented manner. It is relatively easy to convert existing code to use this interface, and it is far easier to read and understand than code with `t_commit` calls scattered around everywhere. `atomic` blocks can be wrapped around existing lock/unlock call pairs or Java `synchronized` blocks. In the latter case, the only difference between the original `synchronized` block and the `atomic` block is that the name of the object variable being used for synchronization is eliminated entirely, thereby reducing the possibility for programmer error. In contrast, with `t_commit` operations, the basic syntax does not offer any clue as to the lock/unlock structure of the code, since transaction beginning and end operations are identical commits. By design, the atomic block structure provides this key insight at a glance.

Just as importantly, regions of parallel code that are *not* accessing shared data, or areas of sequential code, can simply be left unmarked with atomic blocks. This interface therefore provides a clear distinction between code regions where transactions *must* be—the entire `atomic` block must be contained within a single transaction—and where they are optional (between `atomic` blocks they will not hurt, but are not actually necessary). With this model, the programmer no longer needs to bother inserting `t_commit` operations in "nontransactional" parts of the program just to prevent TCC buffer overflow. Instead, the compiler can insert some of these "extra" commits automatically, or even notify the hardware that it is free to commit buffers at any time. If this hardware support is included, the hardware could automatically complete and commit a "transaction" whenever a buffer filled up while in this "nontransactional" mode.

4.2.3 TCC Performance

These TCC programming constructs have been applied to applications explicitly designed to work on conventional parallel systems, and also by using transactions extracted from nominally sequential programs. Simulation was used to evaluate and tune the performance of these applications for large and small-scale TCC systems.

An execution-driven simulator modeled a processor executing at a fixed rate of one instruction per cycle while producing a trace of all transactions in the program. Afterwards, the traces were analyzed on a parameterized TCC system simulator that included 4–32 processors connected by a broadcast network. The TLP-oriented benefits of TCC parallelization are fairly orthogonal to speedup from superscalar ILP extraction techniques within individual processors, so the results may be scaled for systems with processors faster (or slower) than 1.0 instructions per cycle, at least until the commit broadcast bandwidth of the underlying hardware is saturated. Table 4.4 presents the values of three key system parameters selected in order to describe three potential TCC configurations: ideal (infinite bandwidth, zero overheads), single-chip/CMP (high bandwidth, low overheads), and single-board/SMP (medium bandwidth, higher overheads).

Table 4.5 presents the applications used for this study. These applications exhibit a diverse set of concurrency patterns including dense loops (*LUFactor*), sparse loops (*equake*), and task parallelism (*SPECjbb*). C applications were manually modified to use transactions. Java applications ran on a transactional version of the Kaffe JVM [10]. Most were parallelized

TABLE 4.4: Key parameters of our simulations. All cycles are in CPU cycles.

	DESCRIPTION	INTER-CPU BANDWIDTH (BYTES/CYCLE)	COMMIT OVERHEAD (CYCLES)	VIOLATION DELAY (CYCLES)
Ideal	"Perfect" TCC multiprocessor	infinite	0	0
CMP	Realistic multiprocessor, if on a single chip	16	5	0
SMP	Realistic multiprocessor, if on a board	4	25	20

TABLE 4.5: Characteristics of applications used for our analysis.

SOURCE LANGUAGE	BENCHMARK	APPLICATION DESCRIPTION	SOURCE	INPUT	LINES OF CODE	% LINES CHANGED	PRIMARY TCC PARALLELIZATION
Java	Assignment	Resource allocation solver	jBYTEmark	51 × 51 array	556	5.8	Loop: 2 ordered, 9 unordered
	MolDyn	N-body code modeling particles	Java Grande	2048 particles	615	3.3	Loop: 9 unordered
	LUFactor	LU factorization and triangular solve	jBYTEmark	101 × 101 matrix	516	1.9	Loop: 2 ordered, 4 unordered
	RayTrace	3D ray tracer	Java Grande	150 × 150 pixel image	1,233	4.9	Loop: 9 unordered
	SPECjbb	Transaction processing server	SPECjbb	230 iterations w/o random	27,249	1.3	Fork: 5 calls (one per transaction type)
C	art	Image recognition/ neural network	SPEC2000 FP	ref.1	1,270	8.9	Loop: 11 unordered & chunked
	equake	Seismic wave propagation simulation	SPEC2000 FP	ref	1,513	0.8	Loop: 3 unordered
	tomcatv	Vectorized mesh generation	SPEC95 FP	256 × 256	346	2.0	Loop: 7 unordered
	MPEGdecode	Video bitstream decoding	Mediabench	mei16v2.m2v	9,834	4.6	Fork: 1 call

automatically using the Jrpm dynamic compilation system [11], but *SPECjbb* was parallelized by forking transactions for each warehouse task (such as orders and payments). While this benchmark is usually parallelized by assigning separate "warehouses" to each processor, with TCC it was possible to parallelize within a single warehouse, a virtually impossible task with conventional techniques. In all cases, only a modest percentage of the original code needed to be modified to make it parallel.

4.2.3.1 Parallel Performance Tuning

After a programmer divides a program into transactions with the fundamental TCC language constructs described in the last section, most problems that occur will tend to be with performance, and not correctness. In order to get good performance, programmers need to optimize their transactions to balance a few competing goals, which are analogous to those required for good TLS performance: large transactions to minimize overhead, small transactions to avoid buffer overflows and minimize violation losses, maximizing parallel coverage, and minimizing violation-causing dependences between transactions.

Figure 4.7 shows application speedups for successive optimizations on CMP configurations with 4–32 processors, along with a breakdown of how execution time was spent at each step among useful work, waiting for commit, losses to violations, and idling caused by sequential code. The baseline results showed significant speedup, but often needed improvement. Unordered loops were used first, where possible. However, load balancing was rarely a problem in this selection of applications, as exhibited by the low "waiting to commit" times, so this did not improve performance significantly.

The next optimizations were guided by "violation reports," just like those from TLS, that summarized which load–store pairs and data addresses were causing violations, prioritized by the amount of time lost. As with TLS, the guidance provided by these reports can greatly increase programmer productivity by guiding programmers directly to the most critical communication. Violation reports can lead to a wide variety of optimizations, many adapted from traditional parallelization techniques:

Reduction Privatization. As with TLS, associative reduction operations can be privatized on each processor and combined after parallel loops. This should be performed when violations occur due to the sum variable.

Other Variable Privatization. While reduction variables are the most common, other variables may occasionally require privatization. For example, *SPECjbb* was improved by privatization of some shared buffers.

Splitting Transactions into Transactional Groups. Normally, transactions should be monolithic. However, there are times when it is more helpful to break large transactions into two or more

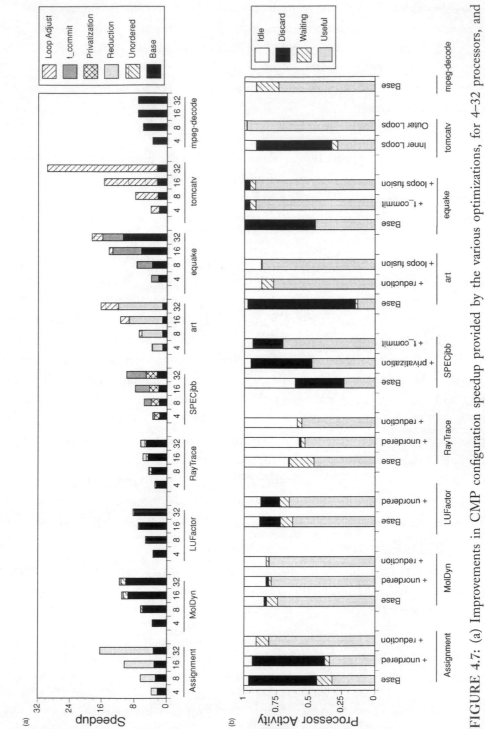

FIGURE 4.7: (a) Improvements in CMP configuration speedup provided by the various optimizations, for 4–32 processors, and (b) processor utilization breakdowns for the 8-processor case.

parts, in a manner analogous to TLS loop slicing. This is performed simply by using a t_commit(0) call to break the transaction into a "transactional group" of two smaller transactions, which the programmer wants to execute sequentially on one processor because they are running highly dependent code from a single thread. Because this operation splits the original transaction into two separate atomic regions, the programmer must ensure that no t_commits are placed in the middle of critical regions that must execute atomically. Despite this limitation, this technique was helpful in solving three different problems. First, inserting a t_commit takes a new rollback checkpoint, limiting the amount of work that can be lost if a violation occurs. This is helpful with *SPECjbb*, with its large transactions and frequent, unavoidable violations. Second, it may be desirable to commit and broadcast changes made by a transaction to other, parallel processors as early as possible. Finally, each commit flushes the processor's TCC write buffer, so judicious t_commits can prevent buffer overflow from large transactions.

Loop Level Adjustment. Different loop nesting levels may be appropriate for different applications. Outer loops provide large granularity, but can sometimes be too large, causing frequent buffer overflows. On the other hand, inner loops may be too small due to startup/commit overheads. Meanwhile, critical loop-carried dependences can occur at any level. *Tomcatv*'s inner loops were small and violation-intensive, forcing the use of outer loops.

Loop Unrolling/Fusion/Fission/Renesting. Any of these common parallelizing compiler tricks can also prove helpful with TCC parallelization. While the techniques are the same, the patterns used are usually somewhat different, with "optimal" transaction sizes being the usual goal.

4.2.3.2 Overall Results

Figure 4.8 presents the best achieved speedups for three configurations of a TCC system (ideal, CMP, SMP) with the number of processors ranging from 4 to 32. For most benchmarks, CMP performance closely tracks ideal performance for all processor counts, but some applications are limited by combinations of unavoidable violations and regions that lacked enough parallel transactions, causing extra processors to idle. Also, *Assignment* and *RayTrace* are limited by large sequential code regions.

The CMP configuration is worse than the ideal one only when insufficient commit bandwidth is available, which only occurred with 32+ processors in this study. In these cases, available bandwidth is insufficient for the amount of committed data, causing processors to stall while waiting for access to the commit network. Similarly, the SMP TCC configuration achieves very little benefit beyond 4–8 processor configurations due to its significantly reduced global bandwidth. From these applications, only *assignment*, *SPECjbb*, and *tomcatv* managed to use the additional processors to significant advantage. This is still promising, however, since

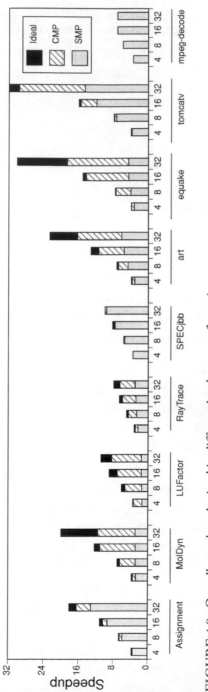

FIGURE 4.8: Overall speedup obtained in different hardware configurations.

online server applications like *SPECjbb* would probably be the most important category of applications to use TCC on systems larger than CMPs.

The most significant TCC hardware is the addition of speculative buffer support. It is critical that the amount of state read and/or written by an average transaction be small enough to be buffered locally. To get an idea about the size of the state storage required, Fig. 4.9 shows the size of the buffers needed to hold the state read (a) or written (b) by 10%, 50%, and 90% of each application's transactions, sorted by the size of the 90% limit, generally in the range of 6–12 KB for read state and 4–8 KB for write state. All applications have a few very large transactions that will overflow, but TCC hardware should be able to achieve good performance as long as this does not occur often.

4.2.3.3 TCC vs. Conventional Parallelization

The previous set of experiments viewed a TCC system as a more scalable version of TLS, which can be used to help parallelize nominally sequential programs using a much simpler parallel programming model. However, it is also possible to use a TCC machine more as a replacement for a conventional shared memory multiprocessor using snoopy cache coherence (SCC). In this case, it is also important to measure the effects of converting code using conventional locks and barriers into code that uses transactions, instead.

Figure 4.10 compares traditional snoopy cache coherence (SCC) with transactional coherence and consistency (TCC) as the number of processors is scaled from 2 to 16, using a selection of SPLASH-2 benchmarks [12] and the *SPECjbb* benchmark used previously. It shows execution time normalized to sequential applications (lower is better). Each TCC bar is broken into five components: *Useful* time spent executing instructions and the TCC API code, *L1 Miss* time spent stalling on loads and stores, *Commit* time spent waiting for the commit token and committing the write set to shared memory, *Idle* time spent idle due to load imbalance, and finally time lost to *Violations*. The SCC bars are slightly different: *Synchronization* is time spent in barriers and locks, while *Communication* is stall time for cache-to-cache transfers. SCC bars do not have violations. Note that the applications are optimized individually for each model.

In general, SCC and TCC perform and scale similarly on most applications up to 16 processors. This demonstrates that continuous transactional execution does not incur a significant performance penalty compared to conventional techniques. Hence, it is worthwhile exploring the advantages it provides for parallel software development. For some applications, as the number of processors increase, time spent in locks and barriers make SCC perform poorly. TCC also loses performance on some applications, but the reasons vary. The differences are generally small, but each application exhibits interesting characteristics:

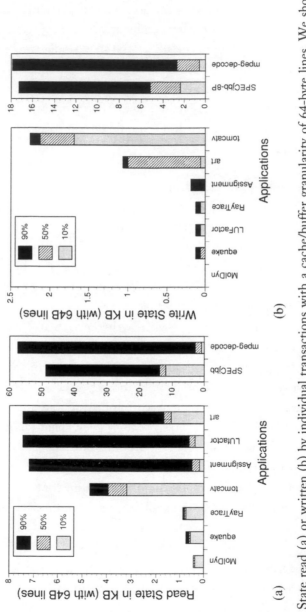

FIGURE 4.9: State read (a) or written (b) by individual transactions with a cache/buffer granularity of 64-byte lines. We show state required by 10%, 50%, and 90% of iterations.

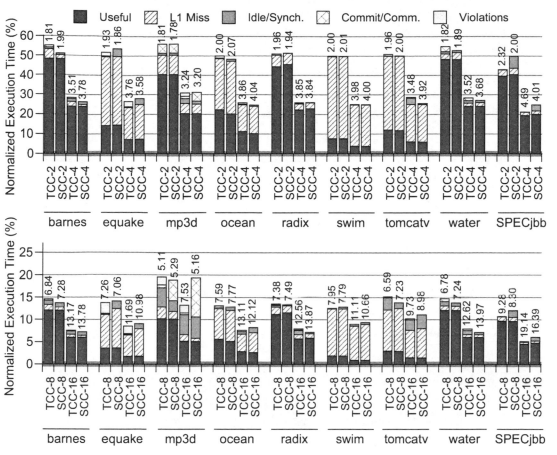

FIGURE 4.10: Normalized execution time for SCC and TCC as we scale the number of processors from 2 to 16. Parallel execution times are normalized to that of a single processor running the original sequential code. The top graph contains runs for 2 and 4 processors while the bottom graph contains the 8 and 16 processor runs; values on the vertical axis change appropriately. Speedups are printed above each bar.

barnes. It scales well on both TCC and SCC, as it only has a small amount of communication between processors. *barnes* has a high operations per word written ratio, which helps TCC amortize the time spent communicating.

equake. The SCC version of *equake* has significant synchronization overhead caused by fine-grained locking to regulate access to a sparse matrix [13]. The TCC version does not require fine-grain lock insertion, but suffers from occasional violations. *equake* is a good example of how the simple TCC programming model provides performance in the face of infrequent sharing.

mp3d. mp3d has a significant amount of communication and false sharing. It is a good example of an irregular parallel program that is difficult to tune. *mp3d* scales well up to four processors on both architectures, but false sharing effects (cache-to-cache transfers) begin to grow at eight processors on SCC. With SCC on 16 processors, cache-to-cache transfers and idle time spent in barriers begin to dominate execution time. In contrast, TCC avoids false sharing by using word-level valid bits and so continues to scale even at 16 processors.

ocean. TCC and SCC perform similarly, but at 16 processors time spent in barriers begins to somewhat hinder scalability on an SCC architecture.

radix. On SCC, *radix* scales well because of its low miss rate and lack of barrier synchronization. The TCC version suffers from load imbalance.

swim. Its execution time is dominated by the high L1 cache miss rate (above 9.15%). *swim* scales similarly with both architectures and with 16 processors is limited by data cache misses that saturate the bus to the L2 cache.

tomcatv. SCC performs better for up to eight processors due to contention for the commit bus: *tomcatv*'s transaction sizes are small and lead to frequent commits. With 16 processors, SCC's performance begins to lag due to synchronization time spent in barriers; TCC's speculative mechanisms avoid some of this delay.

water. water has a tiny miss rate of 0.72%, which ensures that both SCC and TCC scale well to 16 processors. Time stalling for commit poses a small problem for TCC, because the average transaction size is small at 927 instructions and the write state is relatively large at 430 bytes.

SPECjbb. Both TCC and SCC scale well, achieving superlinear speedup due to cache effects. TCC's optimistic concurrency avoids the significant time SCC spends in locks, so TCC achieves better speedups.

Because the SPLASH-2 benchmarks were designed to run well on 1990s-vintage multiprocessors, they did not generally stress the communication resources in either simulated CMP-based system. As a result, interprocessor bandwidths were always fairly low when compared with the requirements for some of the parallelized versions of single-threaded applications described previously.

4.3   MIXING TRANSACTIONAL MEMORY AND CONVENTIONAL SHARED MEMORY

Processor manufacturers and software writers have already invested tremendous amounts of time and money in building conventional shared memory parallel architectures and software optimized for these architectures over the past few decades. As a result, it is likely that processor manufacturers will want to keep this support even as they incorporate transactional memory capabilities into their hardware. Hence, instead of using TCC's simple, 100% transactional model, it is more likely that practical commercial systems will take a hybrid approach that incorporates elements of both transactions *and* conventional shared memory.

From a hardware standpoint, such a hybrid scheme is fairly complex. In addition to the inherently complex conventional shared memory system, memory controllers must also support the simpler transactional protocols. Moreover, the hardware must enforce strict rules regarding how these two separate memory systems interact when accessing the same data in memory to ensure data integrity, prevent deadlock and livelock (such as endless transaction restarts), and so on. Conventional writes can be treated somewhat like very small transactions, but things get trickier when one processor needs to hold a cache line in the exclusive state while another needs to use the same line transactionally, as the two systems treat line ownership differently. These subtle incompatibilities are not insurmountable, but are complex enough that they are beyond the scope of this chapter's discussion. Interested readers should consult our sister lecture [14] for more information on this topic.

On the software side, there are also several important issues. While programs that use all-conventional or all-transactional semantics are straightforward, mixing the two types of semantics in a single program can introduce subtle problems at the transition boundaries that may require extra synchronization or even serialization to avoid. Just as significantly, as we discussed in the last chapter, software engineers generally want a smooth, gradual transition path to using new technologies, so that legacy code may be easily moved to use the new techniques. To give an example of one promising technique, transactional lock removal (TLR) [14, 15] allows most locked critical sections in existing code to be automatically converted to hardware-controlled transactions, while code between critical sections continues to use conventional shared memory protocols. In other words, it automatically transactionalizes locks in existing parallel code in much the same way that TLS can transactionalize loops in existing serial code, with little or no programmer intervention. The performance from this automatic translation is similar to the transactional-vs.-conventional performance comparison results that we have already described: performance is often similar, but in some cases either transactional or conventional semantics will happen to work noticeably better for a particular critical region. Also, some types of critical sections—a prime example is partially overlapping ones—are hard to translate seamlessly into transactions using this model. However, schemes such as TLR offer such a gentle transition path from conventional code that it is likely that the first commercially available transactional memory systems will use similar schemes.

REFERENCES

[1]   L. Hammond, B. D. Carlstrom, V. Wong, M. Chen, C. Kozyrakis, and K. Olukotun, "Transactional coherence and consistency: simplifying parallel hardware and software," *IEEE Micro*, Vol. 24, No. 6, Nov.–Dec. 2004, pp. 92–103.

[2]   L. Hammond, B. Hubbert, M. Siu, M. Prabhu, M. Chen, and K. Olukotun, "The Stanford Hydra CMP," *IEEE Micro.*, Vol. 20, No. 2, Mar.–Apr. 2000, pp. 71–84.

[3]   V. Krishnan and J. Torrellas, "A chip multiprocessor architecture with speculative multithreading," *IEEE Trans. Comput.*, Vol. 48, No. 9, Sept. 1999, pp. 866–880.

[4]   G. Sohi, S. Breach, and T. Vijaykumar, "Multiscalar processors," in *Proc. 22nd Annual Int. Symp. Computer Architecture*, Santa Margherita Ligure, Italy, June 1995, pp. 414–425.

[5]   J. G. Steffan and T. Mowry, "The potential for using thread-level data speculation to facilitate automatic parallelization," in *Proc. 4th Int. Symp. High-Performance Computer Architecture (HPCA-4)*, Las Vegas, NV, Feb. 1998, pp. 2–13.

[6]   J. F. Martinez and J. Torrellas, "Speculative synchronization: applying thread-level speculation to explicitly parallel applications," in *Proc. 10th Int. Conf. Architectural Support for Programming Languages and Operating Systems (ASPLOS-X)*, San Jose, CA, Oct. 2002, pp. 18–29.

[7]   A. Zhai, C. B. Colohan, J. G. Steffan, and T. C. Mowry, "Compiler optimization of scalar value communication between speculative threads," in *Proc. 10th Int. Conf. Architectural Support for Programming Languages and Operating Systems (ASPLOS-X)*, San Jose, CA, Oct. 2002, pp. 171–183.

[8]   T. Sherwood and B. Calder, "Time varying behavior of programs," Dept. of Computer Science and Eng., UCSD, Tech. Rep. No. CS99-630, Aug. 1999.

[9]   M. K. Prabhu and K. Olukotun, "Exposing speculative thread parallelism in SPEC2000," in *Proc. Principles and Practices of Parallel Programming 2005 (PPoPP 05)*, Chicago, IL, June 2005.

[10] T. Wilkinson, Kaffe Virtual Machine, http://kaffe.org, 1997–2002.

[11] M. K. Chen and K. Olukotun, "The Jrpm system for dynamically parallelizing Java programs," in *Proc. 30th Int. Symp. Computer Architecture (ISCA-30)*, San Diego, CA, June 2003, pp. 434–445.

[12] S. Woo, M. Ohara, E. Torrie, J. P. Singh, and A. Gupta, "The SPLASH2 programs: characterization and methodological considerations," in *Proc. 22nd Int. Symp. Computer Architecture (ISCA-22)*, Santa Margherita Ligure, Italy, June 1995, pp. 24–36.

[13] D. O'Hallaron, "Spark98: sparse matrix kernels for shared memory and message passing systems," School of Computer Science, Carnegie Mellon University, Tech. Rep. CMU-CS-97-178, Oct. 1997.

[14] J. R. Larus and R. Rajwar, *Transactional Memory*. San Rafael, CA: Morgan & Claypool Publishers, 2006.

[15] R. Rajwar and J. Goodman, "Transactional lock-free execution of lock-based programs," in *Proc. 10th Int. Conf. Architectural Support for Programming Languages and Operating Systems (ASPLOS)*, San Jose, CA, Oct. 2002, pp. 5–17.

CHAPTER 5

A Multicore World: The Future of CMPs

Along with the many advantages that CMPs provide for software developers, CMPs also have major advantages over conventional uniprocessors for hardware designers. In addition to the power issues that are making them the only viable design option possible in today's processor marketplace, CMPs require only a fairly modest engineering effort for each generation of processors, since each member of a family of processors just requires stamping down a number of copies of the core processor and then making some modifications to the relatively slow logic connecting the processors to tailor the bandwidth and latency of the interconnect with the demands of the processors—but does not necessarily require a complete redesign of the high-speed processor pipeline logic. Over several silicon process generations, the savings in engineering costs can be very significant, and the same engineering effort can be easily amortized across a large family of related processors by simply varying the numbers and clock frequencies of processors to allow essentially the same hardware to function at many different price and performance points. Given the cost of designing a high-performance microprocessor these days, these economies of design are yet another nail in the coffin of conventional, massive uniprocessor systems.

Viewed another way, the transition to CMPs is *inevitable* because past techniques to speed up processor architectures with techniques that do not modify the basic Von Neumann computing model, such as pipelining and superscalar issue, are encountering hard limits. As a result, the microprocessor industry is leading the way to multicore architectures; however, the full benefit of these architectures will not be harnessed until the software industry fully embraces parallel programming. The art of multiprocessor programming, currently mastered by only a small minority of programmers, is more complex than programming uniprocessor machines, and requires an understanding of new computational principles, algorithms, and programming tools. Many techniques are possible on CMPs to ease this transition, but in the end some form of parallel programming will be necessary to fully exploit the potential offered by these systems.

As a result of these trends, chip multiprocessors are taking over the world of general-purpose computing. Throughput computing is the first and most pressing area where CMPs are having an impact. This is because they can improve power/performance results right out of the box, without any software changes, thanks to the large number of independent threads that are available in these already multithreaded applications. In the near future, CMPs should also have an impact in the more common area of latency-critical computations. Companies that produce processors for these markets are already switching to multicore architectures simply because it is too difficult to make more complex uniprocessors, but acceptance of multicore chips in this domain has been slowed by the fact that it is necessary to parallelize most latency-critical software from uniprocessor code into multiple parallel threads of execution, either automatically or manually, in order to really take advantage of chip multiprocessors. CMPs make this process much easier than conventional multiprocessors, due to their short interprocessor communication latencies and ability to implement techniques like the ones presented in this book, but the barrier is visible enough to software that it will continue to be an issue for years to come.

Ultimately, in the future two widely divergent markets for CMPs in general-purpose computers are emerging, based on these two main themes of CMP use. "Manycore" CMPs targeted toward throughput computing, mostly for use in servers, will tend to look a lot like Niagara. These will have a large number of very simple, multithreaded cores highly optimized to maximize performance/Watt while ignoring the potential implications for latency. At the same time, markets whose primary tasks are more latency-sensitive, such as desktop microprocessors, will focus on "multicore" CMPs with fewer moderately superscalar cores. In these systems, single-thread performance will still be a key metric, so the advances in superscalar processor design cannot simply be discarded, even if they do reduce the performance/Watt somewhat. Because multithreaded applications will be developed only very slowly for some of these markets, tools such as TLS and TCC could offer a critical boost to help programmers adapt more readily to the wilderness of parallel programming and thereby accelerate the acceptance of these CMPs for users who have traditionally only purchased uniprocessors. As a result of these trends, desktop processors will no longer simply be "hand-me-down" versions of server processors, since the two families will use radically different design philosophies.

Because of the wide gap between the processors implemented as a result of these two widely divergent design styles, there is some pressure to create hybrids of the two for systems that must execute combinations of throughput-oriented and latency-oriented code. For example, some authors have proposed combinations of both large/fast and small/slow cores on the same die, with the OS responsible for allocating threads among these cores depending upon their latency sensitivity [1]. This combination is advantageous because it provides the "best of both worlds"—a small number of large cores can provide low-latency computation on sequential code when that is necessary, while a larger number of small cores allow more economical execution when sequential computation speed is not essential. Nevertheless, this has not proven to be popular because it would require additional work by software engineers, to design OS schedulers that can intelligently allocate threads among the different processor cores, *and* by hardware engineers, to design and verify the multiple processor cores required for each chip.

In contrast, heterogeneous varieties of CMPs are already being widely developed for the embedded space, since these CMPs can be carefully optimized to run the necessary embedded software applications using precise mixes of simple cores, complex cores, and specialized cores such as DSPs. For example, a typical cell phone contains ASICs that are essentially CMPs with carefully power-optimized combinations of microcontrollers for managing user-interface functions and DSP cores to process the radio signals. At the opposite end of the power scale, the game console market, where the CELL processor [2] is making a big impact following its inclusion in Sony's Playstation 3, is the first place where these more radical hybrids are entering a (somewhat) more general-purpose arena. The CELL incorporates one fairly complex, latency-oriented PowerPC core for executing general-purpose code along with eight simpler cores designed exclusively to accelerate large blocks of throughput-oriented computation for graphics and simulation. To keep the simpler cores as simple as possible, they are not only smaller than the main core, but require code written using a simplified instruction set and coding style. This allows the two types of cores to be highly optimized for their target functions, which is acceptable in the relatively limited space of video game coding. On a completely general-purpose system like a PC, however, this would be less workable.

Hence, two (or more) distinct design chains for general-purpose processors, plus many more families for embedded systems that are optimized in different ways, will become the norm instead of the exception. All of these trends mean that there will be a multitude of widely varying processor designs—and hence space for a wide variety of multicore design techniques to be applied in many ways—in the foreseeable future.

REFERENCES

[1] R. Kumar, D. Tullsen, N. Jouppi, and P. Ranganathan, "Heterogeneous chip multiprocessors," *IEEE Computer*, Vol. 38, No. 11, Nov. 2005, pp. 32–38.

[2] J. A. Kahle, M. N. Day, H. P. Hofstee, C. R. Johns, T. R. Maeurer, and D. Shippy, "Introduction to the CELL multiprocessor," *IBM J. Res. Dev.*, Vol. 49, No. 4/5, pp. 589–604, Sept. 2005.

Author Biography

Kunle Olukotun is a Professor of Electrical Engineering and Computer Science at Stanford University.

Olukotun led the Stanford Hydra project which developed the first chip multiprocessor (multicore chip) with support for thread-level speculation. Using insights gained from the Hydra project, Olukotun founded Afara Websystems to demonstrate the benefits of chip multiprocessor technology for high-throughput, low power server systems. Afara microprocessor technology, called Niagara, was acquired by Sun Microsystems. The Niagara based Sun Fire CoolThreads servers have become one of Sun's fastest ramping products ever.

Olukotun is actively involved in research in computer architecture, parallel programming environments and scalable parallel systems.

Currently, Olukotun directs the Stanford Pervasive Parallelism Lab (PPL) which seeks to proliferate the use of parallelism in all application areas. Olukotun is a Fellow of the ACM. Olukotun received his Ph.D. in Computer Engineering from The University of Michigan.

Lance Hammond: You can find current information on Lance Hammond at his home page, located at http://www.mavam.com/lance/

James Laudon is a Distinguished Engineer with Sun Microsystems. His areas of expertise include multithreading, multiprocessors, and performance modelling. He is currently focused on the architecture of future generations in the UltraSPARC T1 chip multiprocessor line. James joined Sun in July of 2002 through the acquisition of Afara Websystems. At Afara Websystems he managed the architecture and performance team. Prior to Afara, he worked at Broadcom on wired and wireless networking chips, at a superscalar DSP startup, and at Silicon Graphics, where he architected the SGI Origin 2000. James has a B.S. in Electrical Engineering from the University of Wisconsin – Madison and a M.S. and Ph.D. in Electrical Engineering from Stanford University. While at Stanford, James was coarchitect of the Stanford DASH multiprocessor and in his Ph.D. dissertation he proposed Interleaved multithreading, which is the multithreading technique employed in the UltraSPARC T1 chip multiprocessor.